Intelligent and Autonomous

Value Inquiry Book Series

VOLUME 390

The titles published in this series are listed at *brill.com/vibs*

Intelligent and Autonomous

Transforming Values in the Face of Technology

Edited by

Ignas Kalpokas and Julija Kalpokienė

BRILL

LEIDEN | BOSTON

Cover illustration: DALL·E 2 by Open AI.

The Library of Congress Cataloging-in-Publication Data is available online at https://catalog.loc.gov

Typeface for the Latin, Greek, and Cyrillic scripts: "Brill". See and download: brill.com/brill-typeface.

ISSN 0929-8436
ISBN 978-90-04-54722-3 (hardback)
ISBN 978-90-04-54726-1 (e-book)

This book is printed on acid-free paper and produced in a sustainable manner.

Contents

Introduction

Ignas Kalpokas and Julija Kalpokienė

Yes, this is yet another book about digital technologies, artificial intelligence (AI), and the challenges they are bringing to the established ways of life! Surely, one might think, everything that *needs* to be said on the subject must have already been said in the dozens of volumes available on this subject. Nevertheless, this interdisciplinary account of the societal transformations and value challenges resulting from the deployment of 'smart' and increasingly autonomous technologies is intended to reveal the diversity of thinking about the subject matter and the complexity of the challenges that lie ahead. This diversity, however, comes at a cost: a reader expecting to find yet another formulation of *the* definition of AI and *the* concept of human agency vis-à-vis technological artefacts and agents would undoubtedly be disappointed. However, it was a conscious decision of this volume's editors to avoid imposing a totalising view so that the various disciplines and points of view represented in this book are allowed to find their own authentic voice without any restriction or imposition. Hence, the aim of the editors was not to provide a unitary framework but to embrace diversity and show the multiplicity of potential forking paths in researching the relationship between humans and AI broadly conceived. In so doing, the editors hope to enable the readers to think critically and creatively in order to independently interpret the existing research and come to their own conclusions.

While the contributions to this volume are rather diverse, they nevertheless coalesce into a narrative about the current state of the world. In fact, the volume provides an account of a civilisation that is already, for most purposes, cyborg, that is, only definable through the complex relationships between humans and technology. At their most extreme, such relationships can become nothing less than matters of life and death, such as human-AI entanglements within military technologies. As a result, the stakes of AI could not possibly be higher, necessitating proper regulatory frameworks and accountability procedures to be put in place. Simultaneously, though, AI has also become the focus of intense interstate competition on an even broader level – as a general-purpose technology that contributes to inter-state competition as a multiplier of power (both military and economic). In this context, humans occupy an ambivalent position as both a creative force behind new developments in AI and an intellectual resource that states compete for in a manner not dissimilar from other natural resources.

Moreover, as AI encroaches onto ever more spheres of life, matters or regulation and responsibility become crucial, particularly with regards to harm

 | DOI:10.1163/9789004547261_002

being caused by new technologies. The necessity for rules and procedures to be put in place is further underscored by the increasing speed of datafication – with 5G being a notable example – which also means that harms can become instant. No less importantly, dependence on platforms also means dependence on information supplied by them. For this reason, issues pertaining to tackling misinformation and other attempts to abuse our dependency on the digital (and significantly platformised) public sphere deserve particular attention. While formal regulation through law is discussed more often, matters of self-regulation and trust deserve no less attention, which is duly paid in this book. Ultimately, though, this attention of (mis)information has to lead towards a discussion of post-truth. The latter, however, is not seen as some kind of aberration but, instead, as responding to deeply engrained human needs and sustained through daily practice. In the end, though, the high stakes of AI need to be further underscored through its political aspect, opening up the need to further underscore the necessity of politics as a human endeavour vis-à-vis its machinic automation.

Some important caveats are in order, though. When discussing the increasingly intelligent and autonomous nature of today's world, there is a danger to discuss AI and related technologies in a manner not dissimilar from that in which the term 'algorithm' has often been used: as 'sloppy shorthands' that obfuscate more complex processes (for such criticism of discourse on algorithms, see Bogost 2015). Instead, of course, such digital artefacts must be seen as multiple, complex, and perpetually evolving – hence, resisting a stable definition and thus further complicating any attempt to understand this backend of everyday social life (Bucher 2018: 47–48). With regards to algorithms, their prevalence in everyday activities and sheer multiplicity lead towards a condition that Bostrom (2017: 211) identifies as 'algorithmic soup'. In this case, humans are simultaneously affected by multiple digital artefacts of varying complexity and divergent goals, making it hardly possible to isolate one and say that 'the algorithm did this'. Increasingly, the same can be said of AI-powered digital artefacts as well. Nevertheless, stepping out of this digital world becomes hardly possible either as we are faced with a deluge of data, information, and experiences that need to be sifted through for both choice and meaning to become possible (see e.g. Klinger and Swensson 2018). In fact, it would be more surprising if we *did not* end up relying on our digital environment for the sorting of our world (Sunstein 2018: 17), both digital and physical.

However, this sharing of agency raises unavoidable tensions as to the nature of the 'human'. Indeed, as Leslie-McCarthy (2007) insightfully shows in her analysis of Asimov's novels, if there is something to be taken away from visions of the future in science fiction, it is that in a forthcoming society, conundrums involving the constitution of 'personhood' and 'humanity' are going to become unavoidable. In order to reconceptualise agency accordingly, it is useful to shift focus towards affordances, understood as 'the possibilities for action that are available to agents in their environments', but not as pre-given qualities – instead, these are 'aspects of our surroundings that "call upon us" for performing certain action' (Heras-Escribano 2019: 3, 9). Still, the freedom with which such affordances are made use of might be (or become) limited as those failing, unable, or unwilling to take them up would ultimately be left not merely behind but even in the position of a norm transgressor or an offender (Greenfield 2018: 81–82). Hence, matters of the likely impact of technologies and their potential regulation are rendered particularly pressing.

This book uses case analyses and industry insights and blends them with forays into philosophy and ethics in order to conceptualise the mismatch between human values and the values inherent in an increasingly technologized life-world. Bringing together contributors from the disciplines of law, politics, philosophy, and communication studies, this volume develops an interdisciplinary vocabulary for thinking about the questions and antinomies of human-technology interaction while also resisting any deceptively straightforward synthesis.

Structure of the Book

The first chapter of the volume sets the issue of human-machine interaction as more than just the classical 'alignment problem', as the latter would imply that this challenge is ultimately solvable within the existing anthropocentric framework. Instead, a posthumanist lens is applied in order to shed light on the ways in which today's digital technologies, of which AI is at the forefront, displace and destabilise the human as well as the historical, cultural, and political contingencies that had come to constitute the 'anthropos' of anthropocentrism. Hence, a pattern of irresolvable value tension is set up that is then repeated, in different versions and iterations, across the contributions to this volume.

The two following chapters deal with the issues of military AI and the different roles of the humans therein. On the one hand, humans are reduced to potential objects of violence, unleashed by military emerging disruptive technologies, and necessitating an extra degree of protection, particularly in the context of human rights and the laws of war. On the other hand, the same domain can also afford a fundamentally different role to humans vis-à-vis military technology and security thinking – as an active creative agent in the form of talent to be competed for. In this way, a productive tension is set up that provokes further discussion about military-technological (dis)empowerment.

The discussion then moves towards an exploration of human-AI interaction in terms of harm and consent while also exploring matters of agency. The core concern here pertains to a fair balance between the power of humans over AI and the power of AI over humans. Such questions are more pertinent than ever before due to the development of autonomous AI agents as well as 5G and subsequent technologies, intentionally or unintentionally leaving humans out of the loop as to their lived environment and their data. In this way, the matters of regulation transcend the domains of law and politics and become deeply value-laden.

As ever-more power over everyday lives is vested with online platforms, their (self-)regulation becomes a paramount matter of concern. Here, specific attention is turned to social media platforms and the specific matter of disinformation, which significantly challenges human intellectual and cognitive autonomy. Hence, national-and European-level attempts to reinstate the value and the capacity of human intellectual self-determination are explored. Nevertheless, as the contributors demonstrate, such efforts are likely to bring no fewer challenges than they purport to resolve, ultimately identifying a lose-lose situation for both human users and platform infrastructures.

Subsequently, discussion moves to the disruption of politics and decision-making processes more broadly, typically associated with technological advances: post-truth and automated decision-making. Questions as to the specificities of human choice and capacity of self-determination, particularly against the backdrop of datafication and algorithmic governance, are critically examined. Ultimately, the problem is set to revolve around the value of democratic choice and the ethics and feasibility of pre-emptive governance.

Finally, the last chapter of the book raises the question of human obsolescence in the wake of AI and related technologies. It also highlights the problems inherent in establishing and protecting human values (and humans as a value) at the interface of regulation and technology and the indeterminacies facing humanity as a whole in an increasingly technologized future.

References

Bogost, I. (2015, January 15). *The Cathedral of Computation. The Atlantic*, https://www.theatlantic.com/technology/archive/2015/01/the-cathedral-of-computation/384300/.

Bostrom, N. (2017). *Superintelligence: Paths, Data. Strategies*. Oxford and New York: Oxford University Press.

Bucher, T. (2018). *If … Then: Algorithmic Power and Politics*. Oxford and New York: Oxford University Press.

Greenfield, A. (2018). *Radical Technologies: The Design of Everyday Life*. London and New York: Verso.

Heras-Escribano, M. (2019). *The Philosophy of Affordances*. London: Palgrave Macmillan.

Klinger, U. and Swensson, J. (2018). The End of Media Logics? On Algorithms and Agency. *New Media & Society*, 20(12), 4653–4670.

Leslie-McCarthy, S. (2007). Asimov's Posthuman Pharisees: The Letter of the Law Versus the Spirit of the Law in Isaac Asimov's Robot Novels. *Law, Culture and the Humanities*, 3, 398–415.

Sunstein, C. R. (2018). *#Republic*. Princeton and Oxford: Princeton University Press.

CHAPTER 1

Unsustainable Anthropocentrism: Are We Now Posthuman by Design?

Ignas Kalpokas

Today's digital technologies put into question the until recently taken for granted anthropocentric assumptions about the human as a rational, autonomous, and disembodied Cartesian subject. From the anthropocentric perspective, humans had been seen as ontologically, epistemologically, and ethically exceptional and, therefore, entitled to not just to a special standard of care but also to the right of ownership and exploitation of both the biological and the physical environment. Although, it is argued, this view was always both ethically and empirically wrong, it has come to a grinding halt in the face of today's 'smart' digital technologies, particularly those premised upon the surveillance and datafication of everyday life based on increasing adoption of AI tools – to the extent that now humans themselves need to seek shelter from exploitation and technological violence.

The unprecedented level of structuration and datafication of everyday life as well as of the digital (and, increasingly, physical) environment, together with social control technologies and autonomous weapons, underscores not just the futility of the anthropocentric fantasy but also the deep relations of embeddedness that underpin human existence. Hence, this chapter will explore the different ways in which today's digital technologies challenge human autonomy and agency, self-determination, and knowledge. In this, way, it is argued, while true physical cyborgization still lies in the future (experiments with e.g. brain-computer interfaces notwithstanding), one could speak more broadly of the inception of a cyborg society – one which is becoming posthuman by default and the cultural, social, economic, and political strata of which are co-produced by humans and machines.

The argument of this chapter will be developed in three steps. First, posthumanist thinking as such will be discussed, with particular emphasis being put on the violent and exploitative nature of anthropocentrism. Second, transformations that humans undergo as a result of pervasive datafication and algorithmisation of everyday life are analysed, showing how digital technologies are to be seen as explicitly (although not exclusively) posthuman. Finally, a

 | DOI:10.1163/9789004547261_003

cyborgisation thesis is applied at the societal level, demonstrating the ways in which collective human life has been rendered posthuman by default.

1 Posthumanism: Setting the Scene

While much ink has been spilled on discussions of the impact that digital technologies have had (or are going to have) on domains, such as the economy, communication and other forms of social interaction, governance and the law etc., there has still been only scarce attention to the impact they have on the very understanding of the human. What these studies typically assume that, despite the projected transformations, at least in one respect digital technologies are going to continue business as usual – they will continue serving a rational and autonomous human as if nothing has ever happened (or, if more threatening prospects are being raised, such as in warnings about the threats of superintelligence, the conventional anthropocentric standpoint serves as a framework for risk assessment). Hence, the Cartesian human self remains the default standpoint (for a broader discussion, see Kalpokas 2021). Nevertheless, as this chapter aims to demonstrate, there already are ample indications of the ways in which this default setting is changing – and quite fundamentally so – necessitating a new, posthumanist, perspective.

In fact, one should start from the premise that taken-for-granted anthropocentric standpoint has always been both ethically and empirically wrong. The basic premises of the anthropocentric stance are particularly visible in, for example, the central tenets of human development studies. Indeed, current views on human development (and socio-economic developmentality more broadly) acknowledges as good only such human life that has been 'freed from the vicissitudes – the risks and vulnerabilities – of living on the planet, of being part of "nature", of being animal', thereby justifying – and even incentivising – 'exploitation and redesign of nonhuman nature' (Srinivasan and Kasturirangan 2016: 126). The prioritised human is autonomous and disembodied – an ideal Cartesian subject – and this special status, in turn, justifies violence towards the rest of the world. Hence, such understanding of humans legitimises domination and exploitation by emphasising the allegedly exceptional qualities that humans possess that would merit a special standard of care (Srinivasan and Kasturirangan 2016: 127). A more detailed account is provided by Ferrante and Sartori (2016: 176) who identify a three-fold foundation of superiority claims: 'humans are special and privileged entities compared to other living beings (ontology), they are the only sources of knowledge (epistemology) and the sole holders of moral value (ethics)'. Hence, the violence being done is not just physical (actual devastation and degradation) but also symbolic (the denial of standing).

However, it must also be stated that by labelling this thinking 'anthropocentric', one only focuses on a very narrow definition of the *anthropos* – and this is where the empirical wrongfulness of the Cartesian self can be found. Notably, such anthropocentric claim to primacy must be seen as a mere reflection of the interests and ideology of 'a particular subset of humanity that historically has colonized, exploited and plundered both other humans and the non-human world' (De Lucia 2017: 189). Thereby, a hierarchical system is laid bare: one 'imposed upon human beings (intra-species hierarchies) and upon non-human animals and ecosystems (inter-species hierarchies)' (Grear 2015: 230). Even in cases when some value is extended to non-human entities, such as in considerations of animal rights or the emergent debate on robot rights, this is done from a self-arrogated position of power: as Gunkel (2018: 168) observes, 'we project morally relevant properties onto or into those others who we have already decided to treat as being socially significant'. Hence, anthropocentrism here is to be seen as based upon not mere exclusion of non-human but also the vast majority of humanity as well. To put it bluntly, the *anthropos* of anthropocentrism has always been an affluent white Western male who could afford being (or convincingly pretending to be – even to himself) that rational autonomous Cartesian subject *precisely because* the rest was subjected to exploitation for allegedly not making the cut. Everyone and everything else was below in the hierarchy and had to be granted status from above.

It precisely the hierarchical understanding of the organisation of the world that posthumanism strives to overcome (Ferrante and Sartori 2016: 177). However, merely questioning the hierarchical mode of organisation is insufficient, leading to a further questioning of the very foundation of 'the human'. Certainly, displacing 'a single, common standard for "Man" as the measure of all things' (Braidotti 2013: 67; for a similar argument, also see Coeckelbergh 2020: 43), with all of its unavoidably discriminatory appendage (see Braidotti 2019a: 35), is the first and crucial step in this direction. In lieu of such standard, one should embrace what Amoore (2019) calls 'posthuman doubt', a condition of dwelling in the world with only partial or conditional knowledge to lean upon. Likewise, matters of agency and causation must be rethought, replacing the imaginary autonomy of the human self (even in the rethought, more inclusive sense of 'human self') with an account of a more embedded, relational existence – one enmeshed within agglomerations of organic and inorganic nature, data and code, technological artefacts etc. (see e.g. Kalpokas 2021).

The inclusion of digital and technological summands in the above consideration is by no means accidental. Although the taken-for-granted standard of disembodied reason basking in its own autonomous rationality that justifies human exceptionality and dominance over the rest of nature is a quintessential product of modernity, today's digital technologies increasingly displace such views, calling for new categories that would allow a better understanding of the distributed, interembodied, and interrelated presence of humans and other entities (The Onlife Initiative 2015a: 8). In an identical manner, Ess (2015: 89) pronounces the necessity to move from 'the more *individual* sense of rational-autonomous selfhood characteristic of high modern Western thought' to 'more *relational* senses of selfhood'. After all, as Braidotti (2019b: 45) argues from an openly Spinozist perspective, '[w]e are relational beings, defined by the capacity to affect and be affected'. Indeed, it transpires that the self, as an autonomous independent entity, is 'the ultimate abstraction', masking the deep interconnectedness that we have with our surroundings (Oliver 2020: 136). However, while, for example, environmental entanglements could have been self-interestedly ignored for some time (but not indefinitely, as the unfolding climate catastrophe demonstrates), the increasing dependence upon and interoperability with digital technology is becoming immediately evident.

Contrary to the mainstream idea of the environment being passive, submissive to human domination and stewardship, we must understand that humans act on a planet which is 'increasingly hostile and unpredictable' (Burdon 2020: 40) not as some aberration that will pass but, increasingly, as a matter of new normality (Stengers 2015). Hence, the Cartesian disembodiment of the privileged subject, outlined above, is revealed to always have been an *imagined* luxury, one that ignores human embeddedness and relationality at its own peril. To that effect, Chakrabarty (2009) paints an overly soothing picture when arguing for a simple one-way recognition of humans as an ecological force; instead, humans are also acted back upon – with a vengeance. While the climate catastrophe is, in all likelihood, an even more paramount way of the environment pushing back, the rapid shock of the Covid-19 pandemic is perhaps more effective as a watershed moment. Hence, pre-Covidian ideas of freedom and independence-qua-domination are no longer applicable. Instead, as Burdon (2020: 40) stresses, 'morality is not rooted in freedom but in our embeddedness within the Earth system' (see also Hamilton 2016). To that effect, humans must be reframed as merely one part of a holistic approach to the environment (De Vido 2020) which, as argued below, must also extend beyond the natural and into the digital.

Crucially, then, posthumanism is about turning away from the human-nature dualism of Western modernity and, therefore, displacing the human from the status of a privileged subject, instead embedding humans within rich and hybrid agglomerations (Margulies and Bersaglio 2018: 103; see also

Kalpokas 2021). That, among other things, involves acknowledging 'the condition of vulnerability common to all entities' with the explicit aim to 'eradicate injustices caused by human supremacy' and, therefore, necessitating an expansion of current understandings rights and legal personhood (Gellers 2021: 2). With that in mind, one must side with Lupton (2020: 14) in asserting that we are 'more than human' in the sense that 'human bodies/selves are always already distributed phenomena, interembodied with other humans and with nonhumans, multiple and open to the world'. Likewise, then, a privilege position of knowledge must also become questionable. Thus, it is crucial to point out that 'the world is not composed of preexisting and already-formed entities awaiting discovery by human knowers' (Mauthner: 2019: 671; see also Barad 2007) but, instead, dynamic and relational, resisting any stable form of presence. Similarly, following Braidotti (2019b: 47), 'posthuman subjects of knowledge ere embedded, embodied and yet flowing in a web of relations with human and non-human others'. As a result, therefore, knowledge and understanding can only come *from within* the agglomerations that the human subjects are enmeshed in.

The posthumanist critical project is, therefore, aimed at challenging the 'foundational anthropocentric assumptions underpinning Western philosophy and science' that presuppose an autonomous and rational human subject as the exclusive bearer of 'epistemological and moral agency, responsibility, and accountability' (Mauthner 2019: 679). At the heart of posthumanism, therefore, is 'a refusal to bestow ontological priority, primacy, superiority, or separateness on to *any* ontological being, let alone the human' (Mauthner 2019: 679; see also Barad 2007). That is why Braidotti (2019a: 51) ends up calling for a 'transversal alliance [which] today involves non-human agents, technologically-mediated elements, earth-others (land, waters, plants, animals), and non-human inorganic agents (plastic, wires, information highways, algorithms, etc.)'. Hence, a posthumanist ethical stance firmly focuses on the imperative that '[t]he notion of matter as passive and inert, requiring external (human) agency to do anything, is firmly abandoned' (Monforte 2018: 380). Instead, a more equitable relationship is embraced, one that understands the everyday as 'an ongoing composition in which humans and non-humans participate' (Neyland 2019: 11). As a result, therefore, agency must be understood not as a fixed and determinate attribute but, instead, in line with 'ongoing reconfigurings of the world' (Barad 2003: 818). In order to further elucidate such claims, the remaining sections of this chapter will focus on the specific ways in which the technological non-human is making the impossibility of the Cartesian self ever more evident.

2 The Digital Posthuman Condition

As demonstrated in the previous part, posthumanism encourages asking the question 'Do you really know what it means to be human?' (Herbrechter 2013: 38), the answer to which is most certainly indeterminate. While developments such as the climate catastrophe and the Covid-19 pandemic certainly put into question our anthropocentric assumptions and pretences, technological developments must be seen to play a no smaller role in changing the default setting in thinking about the human. In particular, the notions of the human and the self are destabilised by datafication, which leads the construction if decorporealised digital doubles – in other words, the enmeshing of the self with data (Hepp 2020: 159). Likewise, the whole process of datafication also recasts the human from an exploiter of resources towards being a resource. Meanwhile, data and the information (usually algorithmically) derived from them now serve as 'the ultimate moderators of experience and being in today's world' (Herian 2021: 2), shifting the locus of experience from the human mind to data analytics. In this way, the Cartesian *cogito* is externalised and handed over to a machine – or, to be more precise, a cornucopia of networked machines.

To some extent, though, datafication can be seen as a simple continuation of the anthropocentric thrust – at least in its initial conception, it has been a tool for achieving 'a sense of control over the environment and nature' as well as over almost every human self (Herian 2021: 107). However, the end result is quite the opposite – embeddedness of the self within the data-analytic nexus that has become more authoritative in defining the human self than that underlying self can ever be. Moreover, such datafied human self notably lacks finiteness – it is always subject to and redefined by new data and updated results of analytics operations as well as new tools (such as algorithms) introduced by public and private entities holding and analysing the data on the person in question, rendering every such person a permanently beta version of themselves (Kalpokas 2021). In fact, for exactly the same reason, even the digital environment in which individuals find themselves can be likewise described as permanently beta (Hepp 2020: 58–59; see also Kalpokas 2021). Hence, the impossibility of a rational, autonomous, and disembodied Cartesian self is clearly revealed as impossible.

A crucial contributing factor to the undermining of the idea of the independent and self-determining human person has been brought about by what Zuboff (2019) calls 'surveillance capitalism'. Under this mode of socio-economic organisation, humans become data generators in order for that data to be subsequently used to predict individual tastes, choices, and behaviours, not just now, but also into the future – all of that for commercial gain (Zuboff 2019: 8). Crucially, then, '[k]nowledge, authority, and power rest with surveillance capital, for which we are merely "human natural resources"' (Zuboff 2019: 100). In this way, it might be asserted that the digital infrastructure of contemporary societies has undergone a metamorphosis '*from a thing that we have to a thing that has us*' (Zuboff 2019: 202). In an almost identical fashion, Couldry and Mejias (2019: x) describe the currently dominant mode of socio-economic organisation as 'the systematic attempt to turn all human lives and relations into inputs for the generation of profit'. In fact, according to them, the extractive and exploitative practices of contemporary capitalism are comparable to those of the colonial era, only this time it is '*colonization by data*' (Couldry and Mejias 2019: x). In that sense, one could say that the situation both has and has not changed. On the one hand, it is the continuation of that same *anthropos* as the colonialist subject, with the straightforwardly colonial elites having been replaced by technological elites. On the other hand, though, a notable qualitative difference has been the shift from negation of humanness as such, characteristic of colonialism, to co-construction and constant redefinition of humanness, albeit under highly unequal terms.

Human autonomy is clearly revealed as an impossibility in this context. Therefore, as a result of pervasive datafication and algorithmic governance processes, one must think in terms of 'human-data assemblages' that render any distinctions between the human and the non-human indeterminate (Lupton 2020: 14). In other words, it is 'precisely in and through the relations of selves to selves, and selves to others', as these are represented in data, that the algorithmic governance processes of today's world become actionable (Amoore 2020: 8), further underscoring the interactivity and interrelatedness that characterise today's world by default. Again, as per the previous part of this chapter, such an interrelatedness is not a novel invention – at least vis-à-vis the environment, both biological and non-biological, this has always been the case, even if some parts of humanity could have more comfortably ignored that fact than others. However, as datafication involves not just neo-colonisation of the marginalised but also auto-colonisation of the privileged and, no less importantly, permeates and structures even the most intimate aspects of everyday lives, its underscoring of the human lack of autonomy is becoming ever more difficult to ignore.

Further pertaining to the structuration of life by way of datafication, as a result of employment of 'smart' environments and processes of algorithmic governance, any consideration of agency and affordances must involve 'creative interplay of human capabilities and the capacities of more or less smart machines' (Pentzhold and Bischof 2019: 3). Crucially, such '[a]ffordances cannot be determined in advance, but are collectively achieved in interactions between human and technological agents' (Pentzhold and Bischof 2019: 5). Hence, it must be admitted that, as Gherardi (2012: 40) points out, that 'meaning and matter, the social and the technological, are inseparable and they do not have inherently determinable boundaries and properties'. Moreover, we are witnessing the blurring of the distinction between reality and virtuality and between people, nature, and artefacts, including digital and technological ones (The Onlife Initiative 2015b: 44). Crucially, thus, the preceding serves to further illustrate the enmeshing of humans within broader agglomerations and the impossibility of understanding everyday life without its broader contextual conditions in mind – all in stark contrast to abstract disembodied reason by which the *anthropos* of anthropocentrism has been said to operate.

In this 'onlife-world', it transpires, 'artefacts have ceased to be mere machines simply operating according to human instructions' but, instead, are increasingly capable of deciding autonomously, not just learning from their own environments but also generating new environments and personalised experiences, thereby shaping the users and their behaviours from which they were initially only supposed to learn (The Onlife Initiative 2015a: 10). Here, again, Amoore's (2019) focus on doubt gains particular value in expressing 'the many ways in which algorithms dwell within us, just as we too dwell as data and test subjects within their layers' (Amoore 2019: 163). Once again, such shimmering dwelling with and within data and algorithms could not be further away from the self-assertive Cartesian subject on which Western modernity is premised. In an almost identical manner, Lupton and Watson (2020: 4) claims that the relationship between humans and data cannot be seen as static – instead, personal data should be considered as 'lively human-data assemblages that are constantly changing as humans move through their everyday worlds, coming to contact with things such as mobile and wearable devices, online software, apps and sensor-embedded environments'. And while the Cartesian self demands transparency (the latter being a crucial premise for control), a posthumanist self embraces the messiness and relationality that comes with the multiple interminglings of the physical/biological and the digital (for a general discussion of such interminglings, see e.g. Dauvergne 2020).

Moreover, such a non-pre-scriptable (and, hence, doubt-ridden) relationship between humans, data, and computer code makes it possible to think beyond the increasingly prevalent narrative of human-machine competition (Coeckelbergh 2020: 42). Indeed, whether that would be considerations of job loss and technological unemployment or discussions of potential conflicts between the goals of humans and a future AI (with disastrous consequences for humanity), the premise is that of strict separateness between the human and the Other, in this case – the digital-technological Other. This narrative, however, ignores already existing entanglements. Still, such entanglements do not merit the increasingly popular notion of the cyborg (for perhaps the most prominent urtext, see Haraway 1991), primarily because for a blending of the human and the technological (which, after all, is the essence of the cyborg) to make sense, one must first assume a prior entity that is verifiably and unquestionably human. In this sense, the cyborg does not go far enough and remains within the gravitational orbit of Cartesianism. Hence, both competition and blending miss the point. Instead, Amoore's (2019) focus on doubt again serves as a useful counterpoint, the focus here being on multi-way multi-element interactivity without authoritatively known points of reference – or dwelling in the midst of aggregated entanglements.

In particular, AI is worth consideration when discussing posthumanist reconceptualisations. The specialty of AI is that it exceeds tool-like nature by being adaptive and, at least to some extent, autonomous from its own creators, possessing decision-making capacity that is premised upon its own learning processes – in short, AI is 'explicitly posthuman' (Mahon 2018: 80). However, there is also a caveat to the preceding: as evident in much of the futuristic thinking about the relationship between humans and AI, from competition (particularly of a more dystopian kind) to technology-based flourishing, AI is presented in an almost quasi-divine fashion, almost as a deity (malevolent or benevolent) in code (see, perhaps most notably, Geraci 2008; for different peculiarities of the relationship, see also e.g. O'Gieblyn 2017; Brinson 2018; Singler 2020). On the other hand, the focus on ever-present agglomerations might also run the risk of turning into some kind of digital pantheism. Here, Herbrechter's (2013: 3) programmatic requirement '[t]o be able to think the "end of the human" without giving in to apocalyptic mysticism or to new forms of spirituality and transcendence' is particularly apt. Still, it is expected that the present account provides a sufficiently nuanced consideration of the posthuman condition to avoid the mysticism trap.

3 Cyborg Societies: the New Default

The preceding part of this chapter drew to a close with, among other things, a consideration of why cyborgisation is not an appropriate conceptual outlook for analysing the changing conception of the human. However, the situation might be markedly different inasmuch as entire societies are concerned. Here, technology *does* seem to be nor merely co-constituted but also embedded within multiple processes to the effect that key aspects and relationships are revealed (and, indeed, constructed) through human-data-algorithm hybrids. Moreover, as a society is a synthetic composite entity, new members (and new *kinds* of members) are accepted by definition – and in this case, technological artefacts, both physical (devices, sensors, processors, servers) and non-physical (data, code) join in, with at least the same effect as humans.

The first consideration here directly follows from datafication. As quantification, scoring, and ranking become the prevalent ways of status attribution and inferred connections turn into benchmarks for sociability (see, generally, Mau 2019), this hybrid – cyborg – entity properly comes into existence. Perhaps nowhere this has been more manifest than in the various contact tracing apps developed during the Covid-19 pandemic where every contact and interaction, both intentional and accidental, was being evaluated for its risk score. However, the trend is much broader – the drive to assign a numerical score and, therefore, to quantify and evaluate everyone and everything, from university professors to Uber drivers and from one's walking routine to criminality score (as in predictive policing) is a definite sign of rendering societies algorithmically readable and structurable (Mau 2019). In this way, societies become inseparable from their surveillance and analytical infrastructures (see e.g. Andrejevic 2020).

A related process is the intermingling and creation of hybrids through the very process of datafication. For Hepp (2020: 82), we are in the midst of 'processes of an automated, data-based construction of the social world'. Meanwhile, as (Andrejevic 2020: 41) observes, '[w]e live in an epoch of automated mediatization – a time when media technologies and practices are coming to permeate life in unprecedented and unanticipated ways'; these processes, however, are only possible due to 'automated data collection, processing, sorting, and response'. After all, the vision at the heart of surveillance capitalism is 'the everywhere, always-on instrumentation, datafication, connection, communication, and computation of all things, animate and inanimate, and all processes – natural, human, physiological, chemical, machine, administrative, vehicular, financial' effectively refashioning the entire world and its processes as raw data for prediction (Zuboff 2019: 200). Visions of ubiquitous computing, from smart homes to smart cities and smart societies, only reinforce the trend because 'ubiquitous computing is not just a thinking machine; it is an actuating machine designed to produce more certainty *about* us' (Zuboff 2019: 201). This all-inclusive nature of surveillance and datafication is also emphasised by Couldry and Mejias (2019: 8), for whom '[s]ensors never work in isolation but are connected in wider networks that cover ever more of the globe'. Hence, one must keep the global connections and topographies of the digital data-world in mind (see e.g. Dauvergne 2020; Crawford 2021).

In fact, perhaps the only meaningful way to talk about a global society today is to talk about a global data-world, although even in that respect struggles over data and technology are leading to global fracture of different modes and intensities of datafication (see e.g. Lee 2018).

Furthermore, considerable efficiencies can also be achieved when employing digital technologies, particularly those based on AI, in affecting societal-level processes. For one, the use of such technologies to predict choices, tastes, and emotional proclivities, particularly when coupled with the capacity for automatic or semi-automatic content generation, can lead to effectively 'hacking the human brain' by more effectively targeting information and other content (Korinek 2020: 486; see also Ammerman 2019). While most often used for marketing purposes (e.g. targeted advertising), political disinformation campaigns are also a major point of concern (see e.g. Kalpokas and Kalpokiene 2021). A further dimension of societal automation is the use of AI in public administration, predictive policing, or legal judgements (see e.g. Surden 2020). In this case, predictive analytics are employed as tools of future assurance, recomposing society into a sum of individual behaviour pathways extending into the future as guaranteed or, at the very least, statistically highly probable outcomes. In other words, society extends into the future as a multiplicity of technologically derived individual strands of data trends. What connects this with the preceding is the expectation that by foreseeing, tweaking and reweaving those strands as necessary (that 'necessity' again being the result of machinic calculations), future can be optimised for some predefined purpose (improving sales, increasing public safety etc.).

Crucially, then, in today's society, we increasingly cannot treat all the artefacts we encounter simply as tools or instruments – particularly in case of machine learning applications, instructions are not pre-programmed but developed by the system itself from data analysis; hence, 'machine learning systems are deliberately designed to do things that their programmers cannot anticipate, completely control, or answer for', raising further questions of origination, causation, and responsibility (Gunkel 2020: 12). Some artefacts, however, are explicitly posthuman in their function and in the human relationship with them. These primarily include service and social robots that assist the elderly and the children or act as companions to those who are left without human-to-human interaction either by circumstance or by choice; one could also add sex robots to this category (for reasons that probably do not need explicit elaboration) while devices that are less capable in isolation nevertheless are capable of surrounding their users in a posthuman mesh when connected to the Internet of Things (see, generally, Gunkel 2020: 10–11; Liao 2020). No less importantly, technology is challenging aspects that would have been considered quintessentially human, such as such as generation of human-like spoken or written output and, potentially, even creativity, thereby likely to make humans obsolete (see e.g. Gunkel 2020: 8; for more on AI creativity, see e.g. Miller 2019; Du Sautoy 2020).

Increasingly, technological artefacts, both embodied and completely digital, become socially ambivalent, thereby complicating assumptions about humanness, personality, agency, and sociality (Gunkel 2020: 12–13) – and doing so within a two-way relationship that involves not merely objective qualities of artefacts but also their social perception. In other words, such artefacts 'muddy the water [...] by complicating and leaving undecided questions regarding agency, instrumentality, and responsibility' (Gunkel 2020: 37). No less importantly, this ambivalence is also extended through human interaction with the said artefacts. Hence, understanding and regulating such artefacts (e.g. social robots) as simple tools may fail to 'recognize or account for the actual data concerning the way human users respond to, interact with, and conceptualize these devices in practice'; after all, it transpires that '[e]ven when we know that the device is just a "dumb thing", we cannot help but like it and respond to it as another social actor' (Gunkel 2020: 58). With the above in mind, it comes as no surprise that fundamental questions regarding human existence are thrust into the open – after all, '[t]he nonhuman autonomous entity is a necessary counterpoint to the dominant narrative of human identity and meaning', thereby becoming 'our anxious mirror image, our inverted *doppelgänger*' that sets forth the uncanny nature of human mortal existence through its own uncanniness (Kingwell 2020: 340). The uncanniness of the radical other is, in the end, the uncanniness of our own mortal existence.

Digital artefacts, particularly the algorithms involved in governance processes that underlie everything from supply of information to ranking and assessment of humans are also socially significant in their construction of time. As Bucher (2020: 1711) notes, they construct what could be called the 'right-time', which 'is not an experience of a shared 'now' instigated by the temporal regime of the medium but a personalized moment instigated by aggregated individuals, and fuelled by the business models of platforms'. In other words, the time thereby created is based on the premise of delivering *perceived* relevance (ant not necessarily relevance itself) in terms of a 'click' between the user's self-perceived needs and the algorithmic offering at the time when the experience of such a click would be optimal in terms of impact. Hence, we do not necessarily live in a real-time world, but we increasingly live in a right-time world. Hence, while it has become a truism that digital technologies have fundamentally altered space, or at least the social aspects thereof, it now becomes increasingly evident that time is no less affected, leading to a truly Einsteinian moment in the social dimensions of thinking about space-time.

Moreover, the already existing practices of datafication and algorithmic governance are arguably creating the conditions for fundamentally altering the ways in which the socio-political space is organised. The promise is that digital technologies will ultimately enable new *demoi*, unconstrained by geography, communal and administrative boundaries etc. to which people will self-ascribe rather than being born or naturalised into; although that will bring up a further impetus to struggles for access and elimination of digital divides (see e.g. Bernholz, Landemore and Reich 2021: 12) one could also argue that, at least given the way the digital environment is and will likely remain organised, such push towards connectivity is merely a push towards greater datafication and exploitation (Gangadharan 2021: 124; see, more broadly, also Couldry and Mejias 2019). Digital-first update of democracy, whereby technology is employed to manage the conditions for deliberation and choice, ensure the filtering and selection of 'correct' information, create new ways of economic inclusion (such as enabling Universal Basic Income), and bring forth new modes of digital citizenship for more technologically optimised decision-making has been argued to bring forth new emancipatory powers (Ford 2021). Although all of that falls short of suggestions by e.g. Susskind (2018) of completely entrusting governmental functions to an intelligent algorithm (or, at the very least, having different algorithms, and the human teams behind them, running for elections), there definitely is more than a hint of techno-solutionism, the critique of which, formulated a while ago by Morozov (2014), still holds. Such techno-solutionism refers to both technologization of solutions to deeply human problems and doubling down on technological solutions to problems that technology itself has caused and is an important antidote to the thinking that the next clever algorithm or wider implementation of Artificial Intelligence will suddenly solve everything – as if by magic (for a characteristic example of such techno-centric utopias, see e.g. Diamandis and Kotler 2020). In a very similar fashion, for Zuboff (2019: 434), '[c]omputation [...] replaces the political life of the community as the basis for governance'. The outcome is, as succinctly put by Andrejevic (2020: 11), 'the eclipse of politics altogether, insofar as it relies on messy and inefficient forms of deliberation, interpretation, and representation'. In this way, a conflict between rational optimisation and human imperfection becomes clear. In this way, a peculiar substitution takes place: technology becomes the rational autonomous *Anthropos* whereas the human is seen as weak and in need of guidance. To paraphrase Kipling, it becomes 'the AI's burden' to exercise patronising power over humans.

Necessarily for such a machinely optimising framework to emerge, datafied and algorithmified societies outlined above must also, importantly, be predictive societies; Andrejevic (2020: 76–77) rather neatly outlines the scope of such prediction: 'does a data train indicate that someone desires a particular product? Then this can be delivered before they act upon their desire' but also 'Does a pattern of activity indicate the threat of a terrorist attack? A drone can intervene before the threat materializes'. As Berardi (2021: 42) emphasises, these processes bring about a manufactured sense of unavoidability as 'the future is no longer a range of possibilities, but a logically necessary sequence of states of the world', as predicted by trends and correlations spotted in data. This manufactured unavoidability, in turn, leads into '*determinist trap*, a trap in which the possible is captured and reduced to a mere probability, and the probable is enforced as a necessity' (Berardi 2021: 43). Hence, what could be called algorithmic truth becomes the standard of both itself and the society it allegedly represents, rendering the future into only a function of the past and the present.

Seen in this light, the end goal of the wide-ranging deployment of digital technologies, and AI in particular, is 'automating the social' (Andrejevic 2020: 13). In the process, quintessentially human processes, from social interaction to political struggle, are replaced with machinic ones, allegedly in order to eliminate the 'biases, frailties, and idiosyncratic impulses that permeate human judgements – and the cognitive limits that characterize human information processing'; nevertheless, the net result is not necessarily a smoother and more equitable social order (although clearly a more surveilled and datafied one) but, instead, 'the displacement of comprehension by correlation, of explanation by prediction and pre-emption, the triumph of efficiency over other social values' (Andrejevic 2020: 30). The reason for that is relatively straightforward: human and computer reason simply involve different process of decision-making, with the intuitive context-heavy decisions (so far at least) being impossible to codify (Larson 2021) while societies rarely operate along linear 'if … then …' rules. Hence, the posthuman ultimately becomes the default settings in today's societies.

4 Conclusions

This chapter has demonstrated that technologized human life has become posthuman by default. Crucially, technology is not the *cause* of the posthuman condition: anthropocentrism, posthumanism's main object of critique, *has always been* wrong. In this sense, digital technology has simply acted as a mirror that has been held against the lack of mastery and self-sufficiency of the human self. Likewise, the exploitative practices of today;s digital capitalism – one of the fundamental factors underscoring human lack of supremacy in today's digital-first societies – is not about exploitative practices simply emerging out of nowhere but, instead, a complex rearticulation of practices that have existed for centuries. Nevertheless, the central premise, and one that undergirds the thesis of posthumanism as the default setting, is the deep embeddedness of humans within interactive agglomerations that involve biological and non-biological nature, technology and digital elements (data, code etc.) at both micro (individual) and macro (societal) levels. Seen in this light, the subsequent chapters of this volume act as examples of such 'new default' taking hold.

Acknowledgement

Some parts of this research were supported by the EC Pilot Project 'Supporting Collaborative Partnerships for Digital Resilience and Capacity Building in the Times of Disinfodemic (DIGIRES)' LC-01682259.

Bibliography

Ammerman, W. (2019). *The Invisible Brand: Marketing in the Age of Automation, Big Data, and Machine Learning*. New York: McGraw-Hill.

Amoore, L. (2019). Doubt and the Algorithm: On the Partial Accounts of Machine Learning. *Theory, Culture & Society*, 36(6), 147–169.

Amoore, L. (2020). *Cloud Ethics: Algorithms and the Attributes of Ourselves and Others*. Durham (NC): Duke University Press.

Andrejevic, M. (2020). *Automated Media*. London and New York: Routledge.

Barad, K. (2003). Posthumanist Performativity: Toward an Understanding of How Matter Comes to Matter. *Signs*, 28(3), 801–831.

Barad, K. (2007). *Meeting the Universe Halfway: Quantum Physics and the Entanglement of Matter and Meaning*. Durham (NC): Duke University Press.

Berardi, F. (2021). Simulated Replicants Forever? Big Data, Engendered Determinism, and the End of Prophecy. In N. Lushetich (ed.) *Big Data – A New Medium?* London and New York: Routledge, pp. 32–45.

Bernholz, L., Landemore, H. and Reich, R. (2021). Introduction. In L. Bernholz, H. Landemore and R. Reich (eds.) *Digital Technology and Democratic Theory*. Chicago and London: Chicago University Press, pp. 1–22.

Braidotti, R. (2013). *The Posthuman*. Cambridge and Malden: Polity Press.

Braidotti, R. (2019a). A Theoretical Framework for the Critical Posthumanities. *Theory, Culture & Society*, 36(6), 31–61.

Braidotti, R. (2019b). *Posthuman Knowledge*. Cambridge and Medford: Polity.

Brinson, S. (2018, June 27). Dataism: God is in the Algorithm. Medium, https://medium.com/understanding-us/dataism-god-is-in-the-algorithm-84af800205cd.

Bucher, T. (2020). Nothing to Disconnect from? Being Singular Plural in an Age of Machine Learning. *Media, Culture & Society*, doi: 10.1177/0163443720914028.

Burdon, P. D. (2020). Ecological Law in the Anthropocene. *Transnational Legal Theory*, 11(1–2), 33–46.

Chakrabarty, D. (2009). The Climate of History: Four Theses. *Critical Inquiry*, 35(2), 197–222.

Coeckelbergh, M. (2020). *AI Ethics*. Cambridge (MA) and London: The MIT Press.

Couldry, N. and Mejias, U. A. (2019). *The Costs of Connection: How Data Is Colonizing Human Life and Appropriating It for Capitalism*. Stanford: Stanford University Press.

Crawford, K. (2021). *Atlas of AI*. New Haven: Yale University Press.

Dauvergne, P. (2020). *AI in the Wild: Sustainability in the Age of Artificial Intelligence*. Cambridge (MA) and London: The MIT Press.

De Lucia, V. (2017). Beyond Anthropocentrism and Ecocentrism: A Biopolitical Reading of Environmental Law. *Journal of Human Rights and the Environment*, 8(2), 181–202.

De Vido, S. (2020). A Quest for an Eco-Centric Approach to International Law: The Covid-19 Pandemic as a Game Changer. *Jus Cogens*, doi: 10.1007/s42439-020-00031-0.

Diamandis, P. H. and Kotler, S. (2020). *The Future Is Faster Than You Think: How Converging Technologies Are Transforming Business, Industries, and Our Lives*. New York: Simon & Schuster.

Du Sautoy, M. (2020). *The Creativity Code: How AI Is Learning to Write, Paint and Think*. London: 4th Estate.

Ess, Charles (2015). The Onlife Manifesto: Philosophical Backgrounds, Media Usages, and the Futures of Democracy and Equality. In L. Floridi (ed.) *The Onlife Manifesto: Being Human in a Hyperconnected Era*. Cham: Springer, pp. 89–109.

Ferrante, A. and Sartori, D. (2016). From Anthropocentrism to Post-humanism in the Educational Debate. *Relations*, 4(2), 175–194.

Ford, B. (2021). Technologizing Democracy or Democratizing Technology? A Layered-Architecture Perspective on Potentials and Challenges. In L. Bernholz, H. Landemore and R. Reich (eds.) *Digital Technology and Democratic Theory*. Chicago and London: Chicago University Press, pp. 274–308.

Gangadharan, S. P. (2021). Digital Exclusion: A Politics of Exclusion. In L. Bernholz, H. Landemore and R. Reich (eds.) *Digital Technology and Democratic Theory*. Chicago and London: Chicago University Press, pp. 113–140.

Gellers, J. C. (2021). Earth System Law and the Legal Status of Non-Humans in the Anthropocene. *Earth System Governance*, 7, 1–8.

Geraci, R. M. (2008). Apocalyptic AI: Religion and the Promise of Artificial Intelligence. *Journal of the American Academy of Religion*, 76(1), 138–166.

Gherardi, S. (2012). *How to Conduct a Practice-Based Study: Problems and Methods*. Cheltenham and Northampton (MA): Edward Elgar.

Grear, A. (2015). Deconstructing *Anthropos*: A Critical Legal Reflection on 'Anthropocentric' Law and Anthropocene 'Humanity'. *Law Critique*, 26, 225–249.

Gunkel, D. J. (2018). *Robot Rights*. Cambridge (MA) and London: The MIT Press.

Gunkel, D. J. (2020). *How to Survive a Robot Invasion: Rights, Responsibility, and AI*. London and New York: Routledge.

Hamilton, C. (2016). The Anthropocene Rupture. *The Anthropocene Review*, 2(1), 93–106.

Haraway, D. (1991). *Simians, Cyborgs, and Women: The Reinvention of Nature*. London and New York: Routledge.

Hepp, A. (2020). *Deep Mediatization*. London and New York: Routledge.

Herbrechter, S. (2013). *Posthumanism: A Critical Analysis*. London and New York: Bloomsbury.

Herian, R. (2021). *Data: New Trajectories in Law*. London and New York: Routledge.

Kalpokas, I. (2021). *Malleable, Digital, and Posthuman: A Permanently Beta Life*. Bingley: Emerald.

Kalpokas, I. and Kalpokiene, J. (2021). Synthetic Media and Information Warfare: Assessing Potential Threats. In H. Mölder et al. (eds.) *The Russian Federation in the Global Knowledge Warfare*. Cham: Springer Nature, pp. 33–50.

Kingwell, M. (2020). Are Sentient AIs Persons? In M. D. Dubber, F. Pasquale and S. Das (eds.) *The Oxford Handbook of Ethics of AI*. Oxford and New York: Oxford University Press, pp. 324–342.

Korinek, A. (2020). Integrating Ethical Values and Economic Value to Steer Progress in Artificial Intelligence. In M. D. Dubber, F. Pasquale and S. Das (eds.) *The Oxford Handbook of Ethics of AI*. Oxford and New York: Oxford University Press, pp. 475–491.

Larson, E. J. (2021). *The Myth of Artificial Intelligence Why Computers Can't Think the Way We Do*. Cambridge (MA) and London: The Belknap Press of Harvard University Press.

Lee, K. F. (2018). *AI Superpowers: China, Silicon Valley, and the New World*. New York: Houghton, Mifflin, Harcourt.

Liao, S. M. (2020). A Short Introduction to the Ethics of Artificial Intelligence. In S. M. Liao (ed.) *Ethics of Artificial Intelligence*. Oxford and New York: Oxford University Press, pp. 1–42.

Lupton, D. (2020). *Data Selves*. Cambridge and Medford: Polity.

Lupton, D. and Watson, A. (2020). Towards More-than-Human Digital Data Studies: Developing Research-Creation Methods. *Qualitative Research*, doi: 10.1177/1468794120939235.
Mahon, P. (2018). *Posthumanism: A Guide for the Perplexed*. London and New York: Bloomsbury.
Margulies, J. D. and Bersaglio, B. (2018). Furthering Post-Human Political Ecologies. *Geoforum*, 94, 103–106.
Mau, S. (2019). *The Metric Society: On the Quantification of the Social*. Cambridge and Medford: Polity.
Mauthner, N. S. (2019). Toward a Posthumanist Ethics of Qualitative Research in a Big Data Era. *American Behavioral Scientist*, 63(6), 669–698.
Miller, A. I. (2019). *The Artist in the Machine: The World of AI-Powered Creativity*. Cambridge (MA) and London: The MIT Press.
Monforte, J. (2018). What is new for New Materialism for a Newcomer. *Qualitative Research in Sport, Exercise and Health*, 10(3), 378–390.
Morozov, E. (2014). *To Save Everything, Clock Here: The Folly of Technological Solutionism*. New York: Public Affairs.
Neyland, D. (2019). *The Everyday Life of an Algorithm*. London and New York: Palgrave Macmillan.
O'Gieblyn, M. (2017, April 18). God in the Machine: My Strange Journey into Transhumanism. *The Guardian*, https://www.theguardian.com/technology/2017/apr/18/god-in-the-machine-my-strange-journey-into-transhumanism.
Oliver, T. (2020). *The Self Delusion: The Surprising Science of How We Are Connected and why that Matters*. London: Weidenfeld and Nicolson.
Pentzhold, C. and Bischof, A. (2019). Making Affordances Real: Socio-Material Prefiguration, Performed Agency, and Coordinated Activities in Human-Robot Communication. *Social Media + Society*, doi: 10.1177/2056305119865472.
Singler, B. (2020). 'Blessed by the Algorithm': Theistic Conceptions of Artificial Intelligence in Online Discourse. AI & Society, 35, 945–955.
Srinivasan, K. and Kasturirangan, R. (2016). Political Ecology, Development and Human Exceptionalism. *Geoforum*, 75, 125–128.
Stengers, I. (2015). *In Catastrophic Times: Resisting the Coming Barbarism*. London: Open Humanities Press.
Surden, H. (2020). Ethics of AI in Law. In M. D. Dubber, F. Pasquale and S. Das (eds.) *The Oxford Handbook of Ethics of AI*. Oxford and New York: Oxford University Press, pp. 719–736.
Susskind, J. (2018). *Future Politics: Living Together in a World Transformed by Tech*. Oxford and New York: Oxford University Press.
The Onlife Initiative (2015a). The Onlife Manifesto. In L. Floridi (ed.) *The Onlife Manifesto: Being Human in a Hyperconnected Era*. Cham: Springer, pp. 7–13.

The Onlife Initiative (2015b). Background Document: Rethinking Public Spaces in the Digital Transition. In L. Floridi (ed.) *The Onlife Manifesto: Being Human in a Hyperconnected Era*. Cham: Springer, pp. 41–47.

Zuboff, S. (2019). *The Age of Surveillance Capitalism: The Fight for a Human Future at the New Frontier of Power*. London: Profile Books.

CHAPTER 2

Military Emerging Disruptive Technologies: Compliance with International Law and Ethical Standards

Marco Marsili

Margaret Kosal (2019) finds that military applications of emerging disruptive technologies (EDTs)[1] have even greater potential than nuclear weapons to radically change the balance of power. Lethal autonomous weapons systems (LAWS) have been described as the third revolution in warfare, after gunpowder and nuclear arms (Russell, 2015). So far, the debate, stimulated by the Group of Governmental Experts on emerging technologies in the area of lethal autonomous weapons systems (GGE on LAWS) established by the Convention on Prohibitions or Restrictions on the Use of Certain Conventional Weapons (CCW), has focused on artificial intelligence (AI), machine learning (ML), robotics, automation, unmanned autonomous systems (UAS) and semi-autonomous systems (Armin, 2009; Brehm, 2017; Jiménez-Segovia, 2019). Big data, microelectronics, quantum computing/science, cloud computing, deep learning and computer vision algorithms, virtual and augmented reality (VR & AR), 5G networks should also be considered (Marsili, 2021). A broader discussion should include all military EDTs, *inter alia*: space and hypersonic weapons; directed-energy weapons/laser and photonic weapons, just to name a few. So far, a shy approach to the ethical issues of military EDTs is restricted to directed-energy non-lethal weapons (Gordon, 2019).

Though leaders have begun to become aware of legal and ethical implications of the military use of EDTs, these issues still are in the background: security concerns are of pivotal importance (National Security Commission on Artificial Intelligence [NSCAI], 2019; Harrigan, 2019; Reding & Eaton, 2020) and most information and documents are kept confidential, and their circulation is restricted (Bidwell et al., 2018). The compliance of military EDTs with international law (IL), international humanitarian law (IHL), international human rights law (IHRL) and ethical standards should be investigated.[2]

1 For a definition of "disruptive technologies", see: Bower & Christensen (1995).

2 For the purpose of this paper, hereafter we will refer to IL, IHL and IHRL simply as to "IL" as a whole, unless otherwise specified.

 | DOI:10.1163/9789004547261_004

Moving from an historical perspective of the technological innovations of warfare (Pierce, 2004; Payne, 2008), leading researchers discuss the risks of automated killing machines and share their concerns and solutions for reducing risks from intelligent machines. AI and robotics communities face an important ethical decision: whether to support or oppose the development of lethal autonomous weapons systems. Technologies have reached a point at which the deployment of such systems is – practically if not legally – feasible within years, not decades. The war that will flow from the military applications of AI has been termed "intelligent warfare" or "intelligent military technology". The question must be asked as to whether it is intelligent enough to avoid innocent victims and civilian casualties and meet basic legal and human rights standards.

The United States bizarrely finds that autonomous weapons can bring humanitarian benefits. The belief lies in the assumption that reducing the human control over a weapon might increase its accuracy; therefore, surgical use of force would reduce collateral damage and civilian casualties. However, a counterclaim can be proposed: that no warfare will ever be intelligent or humanitarian. Thus far, there are few actors with such technological capabilities and capabilities: US, China and Russia, with the European Union that aims to get a global role in the field of AI, ML and robotics (drones). China is competitive with or ahead of the US., and aims to be the technology leader by 2049 in game-changing fields like hypersonic, quantum sciences, autonomy, artificial intelligence, 5G, genetic engineering and space (Vergun, 2019, Oct. 29). The US Department of Defense (DoD) argues that, while China makes unethical use of AI, the US upholds American values and protects a fundamental belief in liberty and human rights, thus developing principles for using AI in a lawful and ethical manner (Esper, 2019). After all, it is a question of scrutinizing the posture and behaviour of Western allies, which present themselves as the champions of fundamental human rights and civil liberties. On the other hand, we cannot expect that authoritarian regimes, such as Russia and China, respect basic human rights.

Most of Western publications on EDTs – policy and doctrine – are produced by governmental bodies, military institutes and the industry. Obviously, these entities do not question their policy and/or products, which they self-assess to be lawful and ethical but, at least, they consider the issue. An independent assessment should scrutinize this realm. The research on EDTs is mostly linked to governments and not publicised for security reasons, but citizens/voters have the right to be informed promptly about decisions that affect them and on the values at stake. Hence, this paper aims to kick-off an investigation on the ongoing evolution of the US and NATO high-tech policy, strategy and doctrine through extensive document analysis in order to obtain an in-depth understanding and to uncover emerging trends and present and future pitfalls, risks and challenges. The study is based on primary sources that determine the military policy which is checked against legal, ethical, and moral standards examined in the light of current literature. The approach is process-oriented,[3] and is aimed to understand the underlying motives, opinions, and motivations and to provide information about the topic. The purpose is to understand the object of the research through the collection of narrative data, studying the particularities of each document. This contribution does not rest on pre-defined hypotheses, but relies on the ability of the author to divulge meaning from different elements of research without being bound by pre-existing limitations. While this presents a serious challenge, it does open much room for exploration of new fields of research without needing a fixed point of departure – or arrival.

1 The Strategic Advantage of Technological Edge

EDTs give an undisputed geostrategic advantage to those who have them. The Allied Command Transformation (ACT), which operates as the leading agent on innovation for the North Atlantic Treaty Organization, finds that military EDTs have a rapid and major effect on technologies that already exist and disrupt or overturn traditional practices and may revolutionize governmental structures, economies and international security (ACT, 2019). The European Defence Agency (EDA) considers technology a game changer on the defense sector, including military end-users (EDA, 2019). The *Capstone Concept for Joint Operations* (CCJO),[4] the overarching product approved by the US military leadership (Joint Chiefs of Staff [JCS], 2012), that guides the development of future joint capabilities, points out that technology is transforming warfare and reshaping global politics and changing how military and political (civilian) leaders relate.

The US and NATO consider technology to be a competitive advantage. The DoD focuses on innovation and modernization to put innovative technology in the hands of the warfighter (Cronk, 2021). By designing the priorities for twenty first century defense, DoD Secretary Leon Panetta (2012) stated that the Joint Force for the future will be "technologically advanced". Indeed, the

3 For a discussion on "process-oriented" methodologies, see: Onaka, 2013.

4 The current version of the Capstone Concept for Joint Operations: Joint Force 2030 is classified.

CCJO pictures the US armed forces using, *inter alia*, sophisticated technologies in space, cyberspace, and robotics. According to the new defense strategic guidance, advances in machine learning, automated processing and machine-analyst interaction are needed. The Defense Department *Third Offset Strategy*, announced in November 2014 within the *Defense Innovation Initiative*, is aimed to get technological superiority to overcome potential adversaries (Hagel, 2014). This strategy comprises robotics, unmanned autonomous systems,[5] big data and cyber capabilities. The US is determined to achieve more in space, in hypersonics and AI, incorporating space and cyber operations, and integrating them from the seafloor to outer space.

NATO simply copies the US strategy and doctrine: the Warsaw summit communiqué of 2016 calls to apply emerging technologies in the military domain and emphasizes innovation and exploitation of technologies in allied nations; the Brussels Summit declaration of 2018 stresses the importance of maintaining technological edge and advances through innovation. The Alliance is focused on developments in the field of automation, in the integration of AI and the design of UAS capable of operating in multiple domains, and in technological convergence, i.e. the integration of multiple research fields in the identification of the solution to a technological challenge (NATO, 2019). Indeed, EDTs impact conflict management and are key drivers of military operations (Chairman of the Joint Chiefs of Staff [CJCS], 2015, p. 3). Meanwhile, technological advances in hybrid warfare, disruptive technologies and artificial intelligence have created an increasingly complex international security environment (Paxton, 2018).

2 Artificial Intelligence and Machine Learning: the Next Frontier of Warfare?

UAS, AI and ML, hypersonics and directed energy weapons, and 5G networks are game-changing: they are transforming the character and the nature of

5 For the purpose of this paper, we define an autonomous system an anthropogenic system in which technology does not have human influences. Unmanned systems include: air vehicles (unmanned combat air vehicle or UCAV, unmanned aerial vehicle or UAV, unmanned aircraft system or UAS, remotely piloted aerial vehicle or RPAV, and remotely piloted aircraft system or RPAS); unmanned ground vehicles and robotic systems (UGVs); unmanned surface vessels and subsea vehicles, (USV) and unmanned underwater vehicles (UUV); unmanned space vehicles. A UAV, commonly known as "drone", is a component of UAS. The next generation of unmanned systems includes: nuclear powered unmanned submarines; swarming drones; unmanned ships and submersibles.

warfare in radical ways that create enormous challenges and opportunities and will revolutionize the way the military operates in the battlefield (Esper, 2020; Vergun, 2020, 2021).

The research is engaged in embedding cognitive computing in physical spaces (Farrell et al., 2016). Autonomous machines, governed by artificial intelligence, are pushing forward the capacity for space and cyber warfare – AI and ML systems have applications in cyberspace and cyber defense (National Science and Technology Council [NSTC], 2016, p. 36). Military robots will incorporate AI (Cummings, 2017) which will play a greater role in drone data exploitation. The application of ML to AI makes autonomous robots capable to make lethal decisions about humans all on their own. These drones can automatically identify human targets, firing when they choose, with ethical-legal implications that are easily deducible.

The comprehensive book *Machine Learning: An Artificial Intelligence Approach* (Michalski et al., 1983) illustrates how AI and ML are interconnected and often used interchangeably but are not the same thing. Artificial Intelligence is a broader concept than machine learning, which addresses the use of computers to mimic the cognitive functions of humans. First coined in 1956 by John McCarthy, AI is when machines can perform tasks based on algorithms that are characteristic of human intelligence – the idea of AI goes back to at least the 1940s (McCulloch & Pitts, 1943) and was crystalized in Alan Turing's famous 1950 paper, *Computing Machinery and Intelligence*. A report on *One Hundred Year Study on Artificial Intelligence* (Stone et al., 2016) reminds us that there is not precise, universally accepted definition of AI, thus immediately posing the problem of legal definitions – AI includes things like: planning, understanding language, recognizing objects and sounds, learning, and problem solving.

Machine learning is one of the most important technical approaches to AI: it is the ability of machines to receive a set of data and learn for themselves, changing algorithms as they learn more about the information they are processing. Arthur Samuel (1959) defines ML "the ability to learn without being explicitly programmed". Deep learning is one of many approaches to ML, inspired to the structure and function of the brain, i.e. interconnection of neurons, and includes decision learning. This process is based on the use of a series of algorithms – artificial neural networks. In AI/ML, the algorithm is designed to correct/modify itself to perform better in future. That is why we say the AI/ML algorithm is able to learn and has intelligence. Laird proposes a model that combines AI, cognitive science, neuroscience, and robotics (Laird et al., 2017). It must be questioned as to why a system that mimics humans should not make the same mistakes as depicted in the film *War Games* (1983).

The more these technologies advance, the more it is legitimate to question their possible uses. The Organization for Security and Co-operation in Europe (OSCE, 2019) claims that AI is becoming one of the most important technologies of our time. The OSCE warns that possible applications of AI seem almost incomprehensible and its implications for our everyday lives cause both optimistic predictions about future opportunities and serious concerns about potential risks. Schwab (2017) concludes that AI-driven automation is one of the most important economic and social developments in history and has been characterized as the lynchpin of a Fourth Industrial Revolution – the first three industrial revolutions having been driven by steam power, electricity, and electronics.

In 2018, the DoD began to consider seriously that military applications of AI were expected to change the nature of warfare (Ford, 2018). In September 2019, the director of the DoD Joint Artificial Intelligence Center (JAIC), Gen. John N.T. "Jack" Shanahan, predicted that within the next year or two, AI would begin to be employed in warfighting operations and that it would transform warfare (Vergun, 2019; 2020). Gen. Shanahan described a future battle as "algorithms vs. algorithms", with the best algorithm victorious (Vergun, 2019) – that means that without effective AI, the military risks losing the next war. Consequently, it is reasonable to expect an impact of AI also on international humanitarian law, or the law of war, that applies in armed conflicts.

The US is heavily involved in the exploitation of high-tech solutions for military purposes and AI is a strategic priority for the Defense Department (Lopez, 2021a). The application of AI and ML to all-domain information is supposed to give the DoD huge advantages (Garamone, 2021). In a memorandum on the *Establishment of an Algorithmic Warfare Cross-Function Team* (*Project Maven*) of April 26, 2017, the Deputy Secretary of Defense, Robert O. Work, pushes for a more effective integration between AI and ML to maintain advantages over adversaries and competitors. He gathers that DoD should do more efforts to explore the potential of AI, automation, big data, deep learning and computer vision algorithms.

Since 2018, the US military has pushed investments towards autonomous systems driven by AI and ML. The US Army has implemented ML (Leonard, 2018) and, according to the strategic vision set forth by the administration, the ground force is expected to have unmanned vehicles on the battlefield by 2028 (Milley & Esper, 2018). The US Navy warfare is moving from seabed to space, linking manned and unmanned aircraft with satellites, therefore including subsurface, surface, air, space and cyber/EMS in a unique domain of operations (Space and Naval Warfare Systems Command Public Affairs, 2018). The Navy has been reported to place more of its investments toward autonomous systems (Wilkers, 2018) and is testing a Ghost Fleet Overlord Unmanned Surface Vessel (USV), in partnership with the Office of the Secretary of Defense Strategic Capabilities Office (DoD, 2021, June 7). The US National Nuclear Security Administration and the Federal Aviation Administration have developed a counter-UAS that went in a testing phase in June 2018 (The Los Alamos Monitor Online, 2018) while the industry is also pushing ahead (Boeing, 2021).[6]

In March 2021, the Quick Reaction Special Projects program, which is part of the Rapid Reaction Technology Office (RRTO) within the Office of the Under Secretary of Defense for Research and Engineering, published the *2021 Global Needs Statement*, looking for compelling and innovative technologies and ideas in areas involving AI and ML, autonomy, biotechnology, cyber, directed energy, fully networked command, communication. and control, hypersonics, microelectronics, quantum technology, space, use of 5G communications, and other EDTs (DoD, 2021, March 24). The President's DoD budget request for fiscal year 2022 includes the largest-ever research, development, test, and evaluation expenditure and supports modernization to fund such advanced technologies (Under Secretary of Defense, 2021).

We are beyond human-computer interaction, human-robot dialogue, and semiautonomous collaborative systems; we are entering the domain of virtual agents, robots, and other autonomous systems (Ward & DeVault, 2016; Chai et al., 2016). The intensive application of technology leads to the dehumanization of warfare. The impact of AI and autonomous systems on warfare and their employment in a military context gives rise to a debate whether such robots should be allowed to execute missions on their own, especially when human life is at stake (Cummings, 2017). According to the first two Laws of Robotics, laid down by Isaac Asimov, a robot "may not injure a human being or, through inaction, allow a human being to come to harm", and it "must obey the orders given it by human beings except where such orders would conflict" with the previous law.[7] Once again, science fiction has predicted reality, along with the problems of its time.

6 At the NATO Summit held in Brussels in July 2018 the allies agreed to foster innovation, including by further developing partnerships with industry and academia.

7 First Law of Robotics introduced by Isaac Asimov in his 1942 short story *Runaround*, included in *I, Robot* (1950). Near the end of his book *Foundation and Earth* (1986) a zeroth law was introduced: "A robot may not injure humanity, or, by inaction, allow humanity to come to harm".

3 The Legal and Ethical Domain of High-Tech Warfare

The question about the military use of EDTs must be addressed from a legal and ethical point of view. While the legal framework seems to be clear, the same cannot be said of the ethical implications deriving from the military use of certain EDTs, which means considering them as weapons with which to apply the lethal use of force. If, on the one hand, the law lies also on ethical and moral standards, which change in time and space, on the other hand, these concepts continue to be autonomous and are often derived from religious values – which complicates everything. Framed this way, the issue appears to be a game of mirrors – and perhaps it is.

Principles of law are derived from customs and some corps of norms, such as natural and legal rights, human rights, civil rights, and common law. Here arises the question of which source of law has supremacy over the others, whether international or domestic law, which further complicates the framework. Since realizing the impact of military use of AI, that is, around 2016, political and military leaders, as well as scholars, have begun to scrutinize their legality and ethics. Nowadays, these issues are central in the debate on the fairness of this technology. In a media briefing held in August 2019, the director of JAIC, Lt. Gen. Shanahan, acknowledged that the ethical, safe, and lawful use of AI became such a relevant topic that they have to build up a team of specialists, something they never thought about two years before. Gen. Shanahan admitted that in Project Maven "these questions really did not rise to the surface every day, because it was really still humans looking at object detection, classification and tracking" and "[t]here were no weapons involved in that".

The JAIC works with Defense Innovation Board (DIB) and the OSD (Office of the Secretary of Defense) on these questions. The DIB is an independent advisory board set up in 2016 to bring the technological innovation and best practice of Silicon Valley to the US military and provides independent recommendations to the Secretary of Defense. It consists of experts from across commercial sector, research, and academia. The DIB, which supports the research and use of AI as a warfighting tool, called for Defense Department to take the lead in developing ethical AI guidelines within the framework of the *National Defense Strategy* (Vergun, 2019, Nov. 11). Following DIB recommendations, in February 2020 the DoD adopted a series of ethical principles for the lawful use of AI, based on the US Constitution, Title 10 of the US Code, Law of War, existing international treaties and longstanding norms and values.

A DoD memorandum, drafted by Deputy Secretary Kathleen Hicks in May 2021, states that AI could transform the battlefield by increasing the speed of decision making and improving efficiency in back-office operations, but points out that it should be used according to core ethical principles (the so-called "responsible AI", or RAI). The National Security Commission on AI (NSCAI), an independent federal commission established in 2019,[8] is also looking at ethics as part of its broader mandate to study AI and ML for the United States. The *Final Report* stresses the risks associated with AI-enabled and autonomous weapons and raises important legal, ethical, and strategic questions surrounding the use of lethal force, which should be consistent with international humanitarian law (NSCAI, 2021, p. 10). So far, the DoD policy requires a human in the loop within the decision making cycle to authorize the engagement and the use of AI in a Counter-Small Unmanned Aircraft Systems (Lopez, 2021b.).

A seminal report developed in 2016 by the National Science and Technology Council's (NSTC)[9] Subcommittee on Machine Learning and Artificial Intelligence to provides technical and policy advice on topics related to AI, acknowledges that the incorporation of AI in autonomous weapon systems poses policy questions across a range of areas in international relations and security, and leads to concerns about how ensure justice, fairness, and accountability (NSTC, 2016, pp. 1–3, 8). The White House report wonders if the existing regulatory regime is adequate or whether it needs to be adapted (NSTC, 2016, pp. 1–3, 8) and concludes that moving away from direct human control of weapon systems involves some risks and can raise legal and ethical questions (NSTC, 2016, pp. 3, 8, 37). Gordon (2020) suggests that the ethical, sociopolitical, and legal challenges of AI with respect to fundamental rights should be seriously considered.

The report on *Artificial Intelligence and International Affairs* published by Chatham House highlights that the increasing application of AI raises challenges for policymakers and governments (Cummings at al., 2018). The early report on *Artificial Intelligence, Automation and the Economy*, produced in 2016 by the Executive Office of the President of the United States (EOP), warns that AI raises many new policy questions that should be addressed by the Administration (EOP, 2016, p. 45). Cummings at al. (2018) conclude that, under these conditions, new ethical norms are needed. Technology is rule breaker, but rules are necessary.

The upstream question revolves around the ethics of AI and machines, or their ability to make more or less moral decisions, given that the concept of morality is itself elusive. The ethics of AI, a branch of the ethics of technology

8 The NSCAI was established by the *John S. McCain National Defense Authorization Act for Fiscal Year 2019* (P.L. 115–232, § 1051).

9 The NSTC, component of the EOP, is the principal means by which the U.S. Government coordinates science and technology policy.

(or 'technoethics'), is itself a sub-field of ethics, studies the moral behavior of humans as they design, make, use and treat artificially intelligent systems, and the ethics in the behavior of machines (Anderson & Anderson, 2011). The test introduced by Turing in his 1950 paper serves to check if "machines can think" and is an important tool in the philosophical approach to AI. More recently, Coman and Aha (2018), who have explored the ethical implications of AI and assessed its risks, remind us that the ability to say "no" is an essential part of being socio-cognitively human. *Dehumanization of Warfare* (Heintschel von Heinegg et al., 2018), provides a timely overview on autonomous weapons and cyber warfare, considering different new weapon technologies under the same legal rules.

So far, the ethical aspect of technology has been largely scrutinized, albeit limited to AI, ML, robots, and autonomous systems such as drones. In his book *Life 3.0: Being Human in the Age of Artificial Intelligence* (2017), the Swedish-American cosmologist Max Tegmark predicts that there will come a time when machine intelligence reaches the ability to continue to upgrade itself and therefore to advance technologically at an incomprehensible rate independently from humans, thus becoming "superintelligent". This goes far beyond what Müller (2020) has ironically concluded about the major ethical questions stemmed by the unpredictable level of rising of AI and machine learning. Lucas (2004) discusses if there can be an ethical AI but leaves the reader without answer.

Ethics of Artificial Intelligence (Liao, 2020) raises crucial questions of AI and ML. Thes include inquiring into whether autonomous weapon systems are or should be capable of identifying and attacking a target without human intervention or have a greater than human-level moral status? Another question of note involves asking whether we can prevent superintelligent AI from harming us or causing our extinction. Goecke and Rosenthal-von der Pütten (2020) have considered major issues of the current debate on AI and ML from the perspectives of philosophy, theology, and the social sciences. They wonder if machines can replace human scientists and under what condition robots, AI and autonomous systems may have some claim to moral and legal standing and, therefore, can be considered more than a mere instrument of human action. This is a crucial question when we decide to give machines the power to take autonomous decisions – in particular, moral decisions – and to learn by themselves. Tegmark (2017) infers that the risks of AI do not come from malevolence or conscious behavior per se, but rather from the misalignment of the goals of AI with those of humans.

Rébé (2021) argues that AI's physical and decision-making capacities to act on its own implies "a juridical personality", which means that machines should be held accountable for their actions. This consideration opens the door to the accountability of political and military leaders when it comes to the use of lethal force. Lopez Rodriguez et al. (2021) have explored the existing legal concepts of EDTS, *inter alia* AI and UAVS, which are challenging the sovereignty and supremacy of the humans by getting some of their attributes. They wonder if humans should delegate responsibilities to machines and who should be accountable for the decisions. The governance of LAWS, and the fear that they may subvert existing international legal frameworks, are among the main concerns (Maas, 2019). Fairness, accuracy, accountability, and transparency of AI and ML are under discussion (Lo Piano, 2020). Military AI should remain lawful, ethical, stabilizing and safe. Many AI and ML ethics guidelines have been produced, 22 of the major ones have been by analyzed by Hagendorff (2020).

Difficult real-world ethical questions and issues arise from accelerating technological change in the military domain (Allenby, 2015). The critical questions on the legitimacy of drone warfare have been tackled by the scholars through the conceptual lenses of legality and morality (Barela, 2015) without neglecting the political dimension (Galliot, 2015). In a ground-breaking book published in 2009, Krishnan foresees the upcoming introduction to the battlefield of UAVS and the following removal of humans from the battleground. He warns about "the greatest obstacles" of legal and ethical nature which he suggests overcoming through international law. The *Routledge Handbook of War, Law and Technology* (Gow at al., 2019), which provides an interdisciplinary overview of technological change in warfare in the twenty-first century, addresses the challenges international law. The responsibility of military drones under international criminal law is among the main questions raised by military UAVS (Sio and Nucci, 2016), although it remains in the background with respect to ethical and moral considerations for most authors. The use of UAVS impacts on human rights, in particular the right to life. Dos Reis Peron (2014) finds that the employment of drones in targeted killing operations is an indiscriminate and disproportionate use of force.

Drone warfare policy could violate human rights protected under the *International Covenant on Civil and Political Rights* (ICCPR): the right to life; the right to a fair trial; the freedom of association; the right to protection of the family; the right to highest attainable health standards; the right to education; the right of freedom from hunger. In 2014 the UN Human Rights Council (UNHRC) expressed for the very first time serious concern for violations of fundamental human rights in the use of armed drones in military operations (A/HRC/RES/25/22). In a report submitted to the UN General Assembly (UNGA), the special rapporteur on extrajudicial, summary or arbitrary executions, Christof Heyns, highlights the necessity to take precautionary measures to protect the right to life in the use advanced technology and sophisticated weapons, such as drones (A/69/265, §§ 74, 75; see also: UNHRC (2014) A/HRC/26/36, §§ 63, 64). The UN special rapporteur argues that, even if such use cannot be considered inherently unlawful or lawful *per se* (A/69/265, §§ 77, 86) there are serious concerns about the use remote-controlled weapons systems in the military context, which challenge a range of human rights, in particular, the right to life (and bodily integrity in general) and the right to human dignity (A/69/265, §§ 84, 85). Heyns recommends to the international community to adopt a coherent approach in armed conflict and in law enforcement, which covers both international humanitarian law (IHL) and the human rights dimensions, and their use of lethal and less lethal weapons (A/69/265, §§ 89). The UNHRC finds that states must ensure that any measures comply with their obligations under international law, in particular IHL.[10] France, UK and the US are among the six members of the UNHRC which in 2008 voted against the Council Resolution No. 7/7 to ensure the use of armed drones comply with international law. Despite the emerging doctrine on how the future force will operate suggesting minimizing unintended consequences to avoid seriously damaging the international reputation of the United States (CCJO, 2012, p. 7), the *National Defense Strategy 2018*, outlined by DoD Secretary Jim Mattis, concludes that more lethal force is needed in military practices.

In his annual report 2014, the UN special rapporteur on the promotion and protection of human rights and fundamental freedoms while countering terrorism, Ben Emmerson, addresses the legal frame on the employment of armed drones in lethal counter-terrorism operations. The UN special rapporteur calls the HRC to take effective steps, by means of an appropriate resolution aimed at urging all member states to comply with their obligations under international law, including IHL, particularly the principles of precaution, distinction, and proportionality (A/HRC/25/59, § 73). The report complains about a disproportionate number of civilian casualties (A/HRC/25/59, § 21) and suggests fixing, according to the applicable legal principles, some practices and interpretations that "appear to challenge established legal norms" (A/HRC/25/59, § 70).

10 See also: the list of principles concerning the compatibility of anti-terrorism measures with Art. 9 and 10 of the Universal Declaration of Human Rights and Art. 9 and 14 of the ICCPR, in the Report of the Working Group on Arbitrary Detention (A/HRC/10/21, §§ 50–55) drafted by UNHRC chairperson-rapporteur Carmena Castrillo. These principles include that: the detention of persons suspected of terrorist activities shall be accompanied by concrete charges, and in the development of judgments against them, the persons accused shall have a right to the guarantees of a fair trial and the right to appeal.

The international community discusses weather legally binding instruments are necessary to ensure accountability and human control in the use of lethal autonomous weapon systems in armed conflict, or if there is no need for new rules. A 2018 report by the GGE on LAWS, established under the *Convention on Certain Conventional Weapons*, also known as the Inhumane Weapons Convention, wonders if current international law, in particular IHL, fully applies to LAWS (CCW/GGE.2/2018/3). As originally applied only in international armed conflicts, the scope of application of the CCW and its annexed Protocols was broadened in December 2001, in order to extend it to non-international conflicts provided by Art. 2 and 3 common to the Geneva Conventions of 1949, including any situation set forth in Art. 1(4) of Additional Protocol I.[11] The GGE acknowledges that emerging technologies in the area of lethal autonomous weapons pose humanitarian and international security challenges, and affirms that IHL and other applicable international legal obligations apply fully to LAWS, despite the lack of an agreed definition of the terms (CCW/GGE.2/2018/3). The experts agree that the adoption of new weapons, means, or methods of warfare should be consistent with international law and IHL.[12] Suber (2018: 22) suggests determining the term "autonomous", which is still undefined, like many other terms (e.g., outer space, cyberspace, conventional, kinetic, etc.), in order to proceed with the GGE debate.

The question is about the ability of LAWS to distinguish between combatants and civilians and abide by human rights in a world where the understanding of who is a combatant and what is a battlefield in the digital age are quickly overcoming previous standards (CCJO, 2012, p. 3). According to the Final Report of the GGE on LAWS, the capacity of weapons systems to comply with international legal principles of distinction and proportionality, set forth in Art. 36 of Protocol I to the Geneva Conventions, is crucial. A leading researcher like Russell (2015) finds that for current AI systems the principles of necessity, discrimination between combatants and non-combatants, and proportionality between the value of the military objective and the potential for collateral damage, set forth in the 1949 Geneva Convention on humane conduct in war, are difficult or impossible to satisfy. He concludes that, although IHL has no specific provisions for LAWS, it may still be applicable.

The White House report on the future of AI reiterates that LAWS should be incorporated into the US defense planning in accordance with IHL, and that the policy on autonomous weapons must be consistent with shared human

11 Amendment to Art. 1 of the CCW entered into force on May 18, 2004.

12 For a discussion, see: Suber (2018), Weekes and Stauffacher (2018).

values and international and domestic obligations, and must adhere to IHL, including the principles of distinction and proportionality (NSTC, 2016: 3, 37–38). Inconsistencies emerge in US policy, which swings between the defense of fundamental human rights, and the desire to employ LAWS without excessive constraints.[13] The US policy for the fielding of autonomous weapons – systems capable of autonomously selecting and engaging targets with lethal force – requires senior-level DoD approval to be employed (Carter, 2016). The man-machine interface is a focal issue: humans should bear the overall responsibility for coordination and decision-making process. The experts conclude that humans should be held accountable for decisions on the employment of LAWS, and a chain of command and control should be established for their use. A human-centric approach on the human element in the design and (ultimate) decision-making chain when choosing targets, authorising or using (lethal) force.

Some delegations to the CCW GGE on LAWS consider that humans should be held responsible for the final decisions on the use of lethal force. The "accountability approach" considers a set of characteristics related to the functions and type of decisions handed over to machines, which avoids using levels of autonomy related to the loss of human control. Nevertheless, even if some systems have a greater compliance with IHL, machines cannot be programmed to comply with the latter; this responsibility must fall on the human. That is why a minimum level of human control is required, and the question of responsibility is critical.

Selection and engagement of a target is a principal function of LAWS. The ability of the machine for self-learning (without externally fed training data) and self-evolution (without human design inputs) could potentially enable it to redefine targets. The discussion focuses on the spectrum of autonomy and at what exact point on the scale could autonomy become problematic. Some weapon systems, such as drones, have a mix of human decision-making and automation (collaborative systems or human-machine teaming), and others a high level of automation. In the case of LAWS, the validator would be a computer. The importance of maintaining human control and supervision over critical functions of lethal autonomous weapons and the use of lethal force is essential. Experts recommend that LAWS are not anthropomorphized. Cummings argues that the human-machine teaming would be the best solution (CCW/GGE.2/2018/3, p. 16, § 26).

13 See, e.g.: declarations and voting on UNHRC Resolution 7/7 and the 2018 National Defense Strategy.

At the recommendation of the last meeting of the CCW, held in November 2019, 11 guiding principles were adopted related to EDTs in the area of LAWS (CCW/MSP/2019/9). It was affirmed that IL, in particular the UN Charter and IHL, as well as relevant ethical perspectives, should guide the continued work of the Group. The final report concludes, *inter alia*, that: IHL applies fully to LAWS; human-machine interaction in EDTs in the area of LAWS should comply with IL and IHL; human responsibility for decisions cannot be transferred to machines; accountability within a responsible chain of human command and control should be ensured in accordance with IL; the compliance of EDTs and LAWS with IL and IHL and IL should be granted, otherwise new weapons, means or method of warfare must be prohibited.

In a message to the GGE on LAWS, gathered in Geneva for the 2019 meeting, UN Secretary-General António Guterres said that "machines with the power and discretion to take lives without human involvement are politically unacceptable, morally repugnant and should be prohibited by international law". The director of the JAIC closed the discussion by stating that "it would be counterproductive to have outright bans on things that … [nobody] has fully defined" (Shanahan, 2019). So far, the GGE on LAWS has not taken a decision whether calling for a ban on LAWS or not. The debate has reached a deadlock and the 2020 sessions have been postponed indefinitely.

The West strives to combine the strategic advantage resulting from advanced technologies with the main humanitarian principles, which is a difficult task. The OSCE Annual Police Experts Meeting held in 2019 has also addressed key legal, ethical, human rights, and gender-related concerns linked to the application of AI-based technologies in the work of law enforcement authorities and agencies. Opening speakers at the meeting said this technology must be used in strict compliance with human rights and fundamental freedoms (OSCE Secretariat, 2019). OSCE Secretary General Thomas Greminger warned that irresponsible or unethical uses of AI can pose unforeseen risks to liberties and privacy rights, and therefore its use must respect human rights and fundamental freedoms (OSCE Secretariat, 2019). The OSCE Representative on Freedom of the Media (OSCE RFoM), Harlem Désir, concluded that when driven by political or state interests, the use of AI could seriously jeopardize human rights, and called "governments and legislators [to develop] clear and human-rights-friendly policies, including for transparency and accountability, in the use of these technologies" (OSCE RFoM, 2020).

Not even the European Union – most of EU member states, 21 out of 27, are NATO members – has the courage to tackle the issue of AI in warfare. The European Commission (COM/2018/237, COM/2020/64, COM/2020/65) warns that legal and ethical impacts of AI, ML and robotics have to be carefully addressed, but when it comes to propose a draft regulation (COM/2021/205, COM/2021/206), it bans a limited set of uses of AI that contravene EU values or violate fundamental human right but does not include the development and use of AI for military purposes.

4 Conclusions and Recommendations

The major powers focus on science and technology development in order to build military power with strategic impact. High-technology weapons, available also to non-state actors, are assumed to shape the nature of warfare in the twenty-first century. Semiconductors, cloud computing, robotics, and big data are all part of the components needed to develop the AI that will model and define the future battlespace. Artificial intelligence will apply to nuclear, aerospace, aviation, and shipbuilding technologies to provide future combat capabilities. The incorporation of AI into military systems and doctrines will shape the nature of future warfare and, implicitly, will decide the outcome of future conflicts. Before fielding a weapons system, military and political leaders should think about how it can be used and should it be used in a certain manner. A strong and clear regulatory framework is needed.

The tendency of humans is to give more responsibility to machines in collaborative systems. In the future, automatic design and configuration of military operations will be entrusted more and more to the machines. Given human nature, if we recognize the autonomy of machines, we cannot expect anything better from them than the behavior of their creators. So why should we expect a machine to 'do the right thing'? In the light of what has been discussed here, it could be argued that some military applications of EDTs may jeopardize human security.

The total removal of humans from the navigation, command and decision-making processes in the control of unmanned systems, and as such away from participation in hostilities, makes humans obsolete and dehumanizes war. Because of the nature and the technological implications of automated weapons and AI-powered intelligence-gathering tools it is likely that boots on ground will become an exception. A cyber soldier probably will be a human vestige behind the machine.

The rules that will apply to battlespace are unknown. Increased machine autonomy in the use of lethal force raises ethical and moral questions. Is it an autonomous system safe from error? Who will bear the responsibility and accountability for the wrong decision: politicians, low-makers, policy-makes, engineers, or military? Guidelines are needed, and ethical and legal constraints should be considered. Lexicon and definition of terms are also essential, and the international community should find common, undisputed and unambiguous legal formulations. The difference between conventional/unconventional, traditional/non-traditional, kinetic/non-kinetic, and lethal/non-lethal seems to be outdated. A knife, a broken bottle neck, even a fork, a hammer, a baseball bat, or a stone are all unconventional, kinetic, and potentially lethal weapons. Nevertheless, distinguishing between weapons, their effect, and consequence is necessary in order to avoid a cascade effect and undesirable outcomes. The LAWS can lead to an acceleration of a new arms race and proliferation to illegitimate actors such as non-state actors and terrorist groups.

The debate on the application of technology to warfare should cover international law, including IHL, ethics, neuroscience, robotics and computer science. It requires a holistic approach. It is necessary to investigate whether the new domains are actually comparable to the classical ones, and whether current rules are applicable, or if new ones are necessary. Further considerations deriving from the extension of the battlefield to the new domains of warfare concern the use of artificial intelligence in the decision-making process, which, in a fluid security environment, needs to be on target and on time in both the physical and virtual informational spaces. It is not just a legal debate, but also moral and ethical that should be deepened. A multi-disciplinary approach would be useful for designing the employment framework for new warfare technologies.

Acknowledgement

This work received financial support by the European Social Fund (ESF) and by the Fundação para a Ciência e a Tecnologia (FCT), Portugal, under grant SFRH/BD/136170/2018.

References

Allenby, B.R. (2015). The Applied Ethics of Emerging Military and Security Technologies (1st ed.). Routledge. https://doi.org/10.4324/9781315241364.

Allied Command Transformation (ACT). (2019, May 8). *Innovation at the Centre of Allied Command Transformation's Efforts*. https://www.act.nato.int/articles/innovation-centre-allied-command-transformations-efforts.

Anderson, M., & S.L. Anderson (Eds.). (2011). *Machine Ethics*. Cambridge University Press.

Armin, K. (2009). *Killer robots: Legality and ethicality of autonomous weapons*. Ashgate.

Armstrong, P. (2017). *Disruptive Technologies: Understand, Evaluate, Respond.* Kogan Page.

Asimov, I. (1942). Runaround. In *I, Robot* (1950). Gnome Press.

Asimov, I. (1986). *Foundation and Earth*. Doubleday.

Barela, S.J. (2015). Legitimacy and Drones: Investigating the Legality, Morality and Efficacy of UCAVs (1st ed.). Routledge. https://doi.org/10.4324/9781315592152.

Bidwell, C.A. et al. (2018). *Emerging Disruptive Technologies and Their Potential Threat to Strategic Stability and National Security*. Federation of American Scientists. https://fas.org/wp-content/uploads/media/FAS-Emerging-Technologies-Report.pdf.

Boeing. (2021). *Autonomuous Systems. From Seabed-to-Space*. https://www.boeing.com/defense/autonomous-systems/index.page (accessed 2021, June 8).

Bower, J.L., & Christensen, C.M. (1995). Disruptive Technologies: Catching the Wave. *Harvard Business Review*, *73*(1), 43–53.

Brehm, M. (2017). *Defending the Boundary – Constraints and Requirements on the Use of Autonomous Weapon Systems under International Humanitarian and Human Rights Law*. Academy, Briefing No. 9. The Geneva Academy of International Humanitarian Law and Human Rights. https://www.geneva-academy.ch/joomlatools-files/docman-files/Briefing9_interactif.pdf.

Carmena Castrillo, M. (2009). *Report of the Working Group on Arbitrary Detention* (Report No. A/HRC/10/21). UN Human Rights Council (UNHRC).

Carter, A.B. (2012, November 21). *Autonomy in Weapon Systems* (Department of Defense Directive No. 3000.09). https://www.esd.whs.mil/Portals/54/Documents/DD/issuances/dodd/300009p.pdf.

Chai, J.Y., Fang, R., Liu, C., & She, L. (2016). Collaborative Language Grounding Toward Situated Human-Robot Dialogue. *AI Magazine*, *37*(4), 32–45. https://doi.org/10.1609/aimag.v37i4.2684.

Chairman of the Joint Chiefs of Staff (CJCS). (2015, June). *The National Military Strategy of the United States of America 2015*. JCS. https://www.jcs.mil/Portals/36/Documents/Publications/2015_National_Military_Strategy.pdf.

Coman, A., & Aha, D.W. (2018). AI Rebel Agents. *AI Magazine*, *39*(3), 16–26. https://doi.org/10.1609/aimag.v39i3.2762.

Cronk, T.M. (2021, May 13). *Deputy Defense Secretary Travels to Texas With Focus on Innovation, Modernization*. DoD News. https://www.defense.gov/Explore/News/Article/Article/2605954/deputy-defense-secretary-travels-to-texas-with-focus-on-innovation-modernization.

Cummings, M.L. (2017, January). *Artificial Intelligence and the Future of Warfare*. The Royal Institute of International Affairs/Chatham House. https://www.chathamhouse.org/sites/default/files/publications/research/2017-01-26-artificial-intelligence-future-warfare-cummings-final.pdf.

Cummings, M.L., Roff, H.M., Cukier, K., Parakilas, J., & Bryce, H. (2018, June). *Artificial Intelligence and International Affairs: Disruption Anticipated*. The Royal Institute of International Affairs/Chatham House. https://www.chathamhouse.org/sites/default/files/publications/research/2018-06-14-artificial-intelligence-international-affairs-cummings-roff-cukier-parakilas-bryce.pdf.

Dos Reis Peron, A.E. (2014). The 'Surgical' Legitimacy of Drone Strikes? Issues of Sovereignty and Human Rights in the Use of Unmanned Aerial Systems in Pakistan. *Journal of Strategic Security*, *7*(4), 81–93. http://dx.doi.org/10.5038/1944-0472.7.4.6.

Emmerson, Ben. (2014). *Report of the Special Rapporteur on the promotion and protection of human rights and fundamental freedoms while countering terrorism* (Report No. A/HRC/25/59). UN Human Rights Council (UNHRC).

Esper, M.T. (2019, November 5). *Remarks by Secretary Esper at National Security Commission on Artificial Intelligence Public Conference*. DoD. https://www.defense.gov/Newsroom/Transcripts/Transcript/Article/2011960/remarks-by-secretary-esper-at-national-security-commission-on-artificial-intell.

Esper, M.T. (2020, September 16). *Mark T. Esper, Secretary of Defense Engagement at RAND Corporation (Complete Transcript)*. DoD. https://www.defense.gov/Newsroom/Transcripts/Transcript/Article/2351152/secretary-of-defense-engagement-at-rand-corporation-complete-transcript.

European Commission. (2018). *Artificial Intelligence for Europe* (COM/2018/237 final).

European Commission. (2020). *Report on the safety and liability implications of Artificial Intelligence, the Internet of Things and robotics* (COM/2020/64 final).

European Commission. (2020). *White Paper on Artificial Intelligence – A European approach to excellence and trust* (COM/2020/65 final).

European Commission. (2021). *Fostering a European approach to Artificial Intelligence*, (COM/2021/205 final).

European Commission. (2021). *Proposal for a Regulation of the European Parliament and of the Council Laying Down Harmonised Rules on Artificial Intelligence (Artificial Intelligence Act) and Amending Certain Union Legislative Acts* (COM/2021/206 final).

European Defence Agency (EDA). (2019, May 19). *Emerging Game-changers*. https://ec.europa.eu/info/funding-tenders/opportunities/portal/screen/opportunities/topic-details/padr-fddt-emerging-03-2019.

Executive Office of the President of the United States (EOP). (2016, December). *Artificial Intelligence, Automation and the Economy*. https://www.whitehouse.gov/sites/whitehouse.gov/files/images/EMBARGOED AI Economy Report.pdf.

Farrell, R.G. et al. (2016). Symbiotic Cognitive Computing. *AI Magazine*, *37*(3), 81–93. https://doi.org/10.1609/aimag.v37i3.2628.

Ford, C.A. (2018, October 25). *Arms Control and International Security: Why China Technology-Transfer Threats Matter*. https://www.state.gov/t/isn/rls/rm/2018/286889.htm.

Freedman, L. (2017). *The Future of War: A History*. Public Affairs.

Galliott, J. (2015). Military Robots. Mapping the Moral Landscape. Routledge. https://doi.org/10.4324/9781315595443.

Garamone, J. (2021, April 1). *Exercise Reveals Advantages Artificial Intelligence Gives in All-Domain Ops*. DoD News. https://www.defense.gov/Explore/News/Article/Article/2558696/exercise-reveals-advantages-artificial-intelligence-gives-in-all-domain-ops.

Goecke, P.B., & Rosenthal-von der Pütten, A.M. (Eds.). (2020). *Artificial Intelligence. Reflections in Philosophy, Theology, and the Social Sciences.*: Mentis. https://doi.org/10.30965/9783957437488.

Gordon, J. (Ed.). (2020). *Smart Technologies and Fundamental Rights*. Brill Rodopi. https://doi.org/10.1163/9789004437876.

Gow, J., Dijxhoorn, E., Kerr, R., & Verdirame, G. (Eds.). (2019). Routledge Handbook of War, Law and Technology (1st ed.). Routledge. https://doi.org/10.4324/9781315111759.

Group of Governmental Experts on emerging technologies in the area of Lethal Autonomous Weapons Systems (GGE on LAWS). (2018). *Report of the 2018 Group of Governmental Experts on Lethal Autonomous Weapons Systems* (Report No. CCW/GGE.2/2018/3).

Guterres, A. (2019, March 25). *Machines Capable of Taking Lives without Human Involvement Are Unacceptable, Secretary-General Tells Experts on Autonomous Weapons Systems* (SG/SM/19512-DC/3797). UN. https://www.un.org/press/en/2019/sgsm19512.doc.htm.

Hagel, C. (2014, November 15). *Reagan National Defense Forum Keynote*. DoD. https://dod.defense.gov/News/Speeches/Speech-View/Article/606635.

Hagendorff, T. (2020). The Ethics of AI Ethics: An Evaluation of Guidelines. *Minds and Machines 30*(1), p. 99–120. https://doi.org/10.1007/s11023-020-09517-8.

Harrigan, G. (Ed.). (2019). *On the Horizon: Security Challenges at the Nexus of State and Non-State Actors and Emerging/Disruptive Technologies*. NSI. https://nsiteam.com/social/wp-content/uploads/2019/04/DoD_DHS-On-the-Horizon-White-Paper-_FINAL.pdf.

Heintschel von Heinegg, W., Frau, R., & Singer, T. (Eds.). (2018). *Dehumanization of Warfare*. Springer. http://doi.org/10.1007/978-3-319-67266-3.

Heyns, C. (2014). *Report of the Special Rapporteur on extrajudicial, summary or arbitrary executions* (Report No. A/HRC/26/36). UN Human Rights Council (UNHRC).

Heyns, C. (2014). *Extrajudicial, summary or arbitrary executions* (Report No. A/69/265). UN General Assembly (UNGA).

Hicks, K. (2021, May 26). *Memorandum for Senior Pentagon Leadership on Implementing Responsible Artificial Intelligence in the Department of Defense*. https://media.defense.gov/2021/May/27/2002730593/-1/-1/0/IMPLEMENTING-RESPONSIBLE-ARTIFICIAL-INTELLIGENCE-IN-THE-DEPARTMENT-OF-DEFENSE.PDF.

Jiménez-Segovia, R. (2019). Autonomous weapon systems in the Convention on certain conventional weapons: Legal and ethical shadows of an autonomy, under human control? REEI, 37, 1–33.

Joint Chiefs of Staff (JCS). (2012, September). *Capstone Concept for Joint Operations: Joint Force 2020* (CCJO). JCS. https://www.benning.army.mil/mssp/security%20topics/Global%20and%20Regional%20Security/content/pdf/CCJO%20Joint%20Force%202020%2010%20Sept%2012.pdf.

Kelsen, H. (1949). *General Theory of Law and State*. Harvard University Press.

Kosal, M.E. (Ed.). (2019). *Disruptive and Game Changing Technologies in Modern Warfare*. Springer.

Krishnan, A. (2009). Killer Robots: Legality and Ethicality of Autonomous Weapons (1st ed.). Routledge. https://doi.org/10.4324/9781315591070.

Laird, J.E., Lebiere, C., & Rosenbloom, P.S. (2017). A Standard Model of the Mind: Toward a Common Computational Framework across Artificial Intelligence, Cognitive Science, Neuroscience, and Robotics. *AI Magazine*, *38*(4), 13–26. https://doi.org/10.1609/aimag.v38i4.2744.

Leonard, M. (2018, July 2). *Army leverages machine learning to predict component failure*. Defense Systems. https://defensesystems.com/articles/2018/07/03/army-vehicle-predictive-maintenance.aspx.

Liao, S.M. (2020) (Ed.). *Ethics of Artificial Intelligence*. Oxford University Press. 10.1093/oso/9780190905033.001.0001.

Lopez, C.T. (2021a, March 23). *If DOD Wants AI in Its Future, It Must Start Now*. DoD News. https://www.defense.gov/Explore/News/Article/Article/2547622/if-dod-wants-ai-in-its-future-it-must-start-now.

Lopez, C.T. (2021b, February 3). *Defense Official Discusses Unmanned Aircraft Systems, Human Decision-Making, AI*. DoD News. https://www.defense.gov/Explore/News/Article/Article/2491512/dod-official-discusses-unmanned-aircraft-systems-human-decision-making-ai.

Lopez Rodriguez, A.M., Green, M.D., & Kubica, M.L. (2021). *Legal Challenges in the New Digital Age*. Brill Nijhoff.

Lo Piano, S. (2020). Ethical principles in machine learning and artificial intelligence: cases from the field and possible ways forward. *Humanit Soc Sci Commun*, *7*(9). https://doi.org/10.1057/s41599-020-0501-9.

Lucas, R. (2004). An Outline for Determining the Ethics of Artificial Intelligence. *Manusya: Journal of Humanities*, *7*(4), 68–82. https://doi.org/10.1163/26659077-00704006.

Maas M.M. (2019). Innovation-Proof Global Governance for Military Artificial Intelligence? *Journal of International Humanitarian Legal Studies*, *10*(1), 129–157. https://doi.org/10.1163/18781527-01001006.

Marsili, M. (2021). Epidermal Systems and Virtual Reality: Emerging Disruptive Technology for Military Applications. *Key Engineering Materials*, *893*, 933–101.

Mattis, J.N. (2018). *Remarks by Secretary Mattis on National Defense Strategy*. https://dod.defense.gov/News/Transcripts/Transcript-View/Article/1702965/remarks-by-secretary-mattis-on-national-defense-strategy.

Mattis, J.N. (2018). *2018 National Defense Strategy*. U.S. Department of Defense.

McCarthy, J., Minsky, M., Rochester, N., & Shannon, C. (2006). A Proposal for the Dartmouth Summer Research Project on Artificial Intelligence, Aug. 31, 1955. *AI Magazine*, *27*(4), 12–14. https://doi.org/10.1609/aimag.v27i4.1904.

McCulloch, W.S., & and Pitts, W.H. (1943). A Logical Calculus of the Ideas Immanent in Nervous Activity. *Bulletin of Mathematical Biophysics*, *5*(43), 115–133.

Meeting of the High Contracting Parties to the Convention on Prohibitions or Restrictions on the Use of Certain Conventional Weapons Which May Be Deemed to Be Excessively Injurious or to Have Indiscriminate Effects. (2019). *Final report* (Report No. CCW/MSP/2019/9). UN.

Michalski, R.S., Carbonell, J., & Mitchell, T.M. (Eds.). (1983). *Machine Learning: An Artificial Intelligence Approach*. Tioga.

Milley, M.A., & Esper, M.T. (2018, June 5). *The Army Vision*. U.S. Army. https://www.army.mil/e2/downloads/rv7/vision/the_army_vision.pdf.

Müller, V.C. (2020). Ethics of Artificial Intelligence and Robotics. In E.N. Zalta (Ed.), *The Stanford Encyclopedia of Philosophy* (Winter Edition), https://plato.stanford.edu/archives/win2020/entries/ethics-ai.

National Security Commission on Artificial Intelligence (NSCAI). (2019, July). *Initial Report*. https://www.nscai.gov/wp-content/uploads/2021/01/NSCAI_Initial-Report-to-Congress_July-2019.pdf.

National Science and Technology Council (NSTC). (2016, October). *Preparing for the Future of Artificial Intelligence*. The White House. https://obamawhitehouse.archives.gov/sites/default/files/whitehouse_files/microsites/ostp/NSTC/preparing_for_the_future_of_ai.pdf.

National Security Commission on Artificial Intelligence (NSCAI). (2021, March). *Final Report, Executive Summary*. https://www.nscai.gov/wp-content/uploads/2021/03/Final_Report_Executive_Summary.pdf.

NATO Heads of State and Government. (2016, July 9). *Warsaw Summit Communiqué*. Press release 100. https://www.nato.int/cps/en/natohq/official_texts_133169.htm.

NATO Heads of State and Government. (2018, July 11). *Brussels Summit Declaration*. Press release 74. https://www.nato.int/cps/en/natohq/official_texts_156624.htm.

NATO (2019, October 4). *NATO focuses on future of advanced technologies*. https://www.nato.int/cps/en/natohq/news_169419.htm?selectedLocale=en.

Onaka, F. (2013). Aspects of Process Theories and Process-Oriented Methodologies in Historical and Comparative Sociology: An Introduction. *Historical Social Research/*

Historische Sozialforschung, 38(2), 161–171. https://doi.org/10.12759/hsr.38.2013.2.161-171.

Organization for Security and Co-operation in Europe (OSCE) (n.d.). *2019 OSCE Annual Police Experts Meeting: Artificial Intelligence and Law Enforcement – An Ally or Adversary?*. https://www.osce.org/event/2019-annual-police-experts-meeting.

Organization for Security and Co-operation in Europe (OSCE) Secretariat. (2019, September 23). *Law enforcement agencies should embrace Artificial Intelligence to enhance their efficiency and effectiveness, say police experts at OSCE meeting*. https://www.osce.org/chairmanship/432152.

OSCE Representative on Freedom of the Media (OSCE RFoM). (2020, March 10). *OSCE roundtable and expert meeting in Vienna – the impact of artificial intelligence on freedom of expression*. http://www.osce.org/representative-on-freedom-of-media/448225.

OSCE Representative on Freedom of the Media (OSCE RFoM). (2020, March). *Spotlight on Artificial Intelligence & Freedom of Expression. OSCE RFoM Non-paper on the Impact of Artificial Intelligence on Freedom of Expression. #SAIFE*. https://www.osce.org/representative-on-freedom-of-media/447829?download=true.

Panetta, L.E. (2012, January 5). *Sustaining U.S. Global Leadership: Priorities for 21st Century Defense*. DoD. https://archive.defense.gov/news/defense_strategic_guidance.pdf.

Paxton, J. (2018, November 16). *Trident Juncture and the information environment*. NATO Review. https://www.nato.int/docu/review/articles/2018/11/16/trident-juncture-and-the-information-environment/index.html.

Payne, K. (2018). *Strategy, Evolution, and War: From Apes to Artificial Intelligence*. Georgetown University Press.

Pierce, T.C. (2004). *Warfighting and Disruptive Technologies: Disguising Innovation*. Routledge.

Protocol Additional to the Geneva Conventions of 12 August 1949, and relating to the Protection of Victims of International Armed Conflicts (Protocol 1). Treaty Series, 1125, 3.

Rébé, N. (2021). *Artificial Intelligence: Robot Law, Policy and Ethics* (Nijhoff Law Specials, Vol. 102). Brill Nijhoff. https://doi.org/10.1163/9789004458109.

Reding, D.F., & Eaton, J. (2020, March). *Science & Technology Trends: 2020–2040. Exploring the S&T Edge*. NATO Science & Technology Organization, Office of the Chief Scientist. https://www.nato.int/nato_static_fl2014/assets/pdf/2020/4/pdf/190422-ST_Tech_Trends_Report_2020-2040.pdf.

Russell, S. et al. (2015). Robotics: Ethics of artificial intelligence. *Nature, 521*(7553), p. 415–418. https://doi.org/10.1038/521415a.

Samuel, A.L. (1959). Some Studies in Machine Learning Using the Game of Checkers. *IBM Journal of Research and Development, 44*(1.2), 210–229.

Schwab, K. (2017). *The Fourth Industrial Revolution*. Crow Business.

Shanahan, J. (2019, August 30). *Lt. Gen. Jack Shanahan Media Briefing on AI-Related Initiatives within the Department of Defense*. DoD. https://www.defense.gov/Newsroom/Transcripts/Transcript/Article/1949362/lt-gen-jack-shanahan-media-briefing-on-ai-related-initiatives-within-the-depart.

Sio, F.S.D., & Nucci, E.D. (2016). Drones and Responsibility: Legal, Philosophical and Socio-Technical Perspectives on Remotely Controlled Weapons (1st ed.). Routledge. https://doi.org/10.4324/9781315578187.

Space and Naval Warfare Systems Command Public Affairs. (2018, January-March). Seabed to Space: US Navy Information Warfare Enriches Cyber Resiliency, Strategic Competition. *CHIPS*. http://www.doncio.navy.mil/CHIPS/ArticleDetails.aspx?ID=10018.

Stone, P. et al. (2016, September). "Artificial Intelligence and Life in 2030". *One Hundred Year Study on Artificial Intelligence: Report of the 2015–2016 Study Panel*. Stanford University. http://ai100.stanford.edu/2016-report.

Suber, R. (2018, February). *Artificial Intelligence: Autonomous Technology (AT), Lethal Autonomous Weapons Systems (LAWS) and Peace Time Threats*. CT4Peace Foundation Zurich. https://ict4peace.org/wp-content/uploads/2018/02/2018_RSurber_AI-AT-LAWS-Peace-Time-Threats_final.pdf.

Tegmark, M. (2017). *Life 3.0. Being Human in the Age of Artificial Intelligence*. Alfred Knopf.

The Los Alamos Monitor Online. (2018, June 13). *LANL designates restricted airspace where unauthorized drone flights prohibited, including additional 'No Drone Zone'*. https://www.lamonitor.com/content/lanl-designates-restricted-airspace-where-unauthorized-drone-flights-prohibited-including-ad.

Turing, A.M. (1959). Computing Machinery and Intelligence. *Mind*, *59*(236), 433–460. https://doi.org/10.1093/mind/LIX.236.433.

Under Secretary of Defense (Comptroller)/Chief Financial Officer (2021). DoD Budget Request. Defense Budget Materials-FY2022. DoD. https://comptroller.defense.gov/Budget.

UN Convention on Prohibitions or Restrictions on the Use of Certain Conventional Weapons Which May Be Deemed to Be Excessively Injurious or to Have Indiscriminate Effects. *Treaty Series*, *1342*, 137.

UN General Assembly (UNGA). (1966). International Covenant on Civil and Political Rights. *Treaty Series*, *999*, 171.

UN Human Rights Council (UNHRC). (2014). *Ensuring use of remotely piloted aircraft or armed drones in counterterrorism and military operations in accordance with international law, including international human rights and humanitarian law* (Resolution No. A/HRC/RES/25/22).

UN Human Rights Council (UNHRC). (2008). *Protection of human rights and fundamental freedoms while countering terrorism* (Resolution No. 7/7).

U.S. Department of Defense. (2020, February 24). DOD Adopts Ethical Principles for Artificial Intelligence. https://www.defense.gov/Newsroom/Releases/Release/Article/2091996/dod-adopts-ethical-principles-for-artificial-intelligence.

U.S. Department of Defense. (2021, March 24). *2021 Global Needs Statement* (RRTO-20210324-W-Global). https://sam.gov/opp/e67a3de6522b46cd8afd09a58f17ebe2/view.

U.S. Department of Defense. (2021, June 7). *Ghost Fleet Overlord Unmanned Surface Vessel Program Completes Second Autonomous Transit to the Pacific*. https://www.defense.gov/Newsroom/Releases/Release/Article/2647818/ghost-fleet-overlord-unmanned-surface-vessel-program-completes-second-autonomou.

Vergun, D. (2019, September 24). *AI to Give U.S. Battlefield Advantages, General Says*. https://www.defense.gov/explore/story/Article/1969575/ai-to-give-us-battlefield-advantages-general-says.

Vergun, D. (2019, October 29). *Chinese Set Sights on High-Tech Production*. DoD. https://www.defense.gov/explore/story/Article/2002618/chinese-set-sights-on-high-tech-product-dominance.

Vergun, D. (2019, November 1). *Defense Innovation Board Recommends AI Ethical Guidelines*. DoD News. https://www.defense.gov/explore/story/Article/2006646/defense-innovation-board-recommends-ai-ethical-guidelines.

Vergun, D. (2019, November 5). Without Effective AI, Military Risks Losing Next War, Says AI Director. https://www.defense.gov/explore/story/Article/2009288/without-effective-ai-military-risks-losing-next-war-says-ai-director.

Vergun, D. (2020, November 10). *DOD Works With Industry to Ramp Up 5G Network Innovation*. DoD News. https://www.defense.gov/Explore/News/Article/Article/2410280/dod-working-with-industry-to-ramp-up-5g-network-innovation.

Vergun, D. (2020, June 5). *Experts Predict Artificial Intelligence Will Transform Warfare*. DoD News. https://www.defense.gov/Explore/News/Article/Article/2209480/experts-predict-artificial-intelligence-will-transform-warfare.

Vergun, D. (2021, May 13). *Speed, Integration Needed to Deter China, Russia, Vice Chairman Says*. DoD News. https://www.defense.gov/Explore/News/Article/Article/2606552/speed-integration-needed-to-deter-china-russia-vice-chairman-says.

Ward, N.G., & DeVault, D. (2016). Challenges in Building Highly-Interactive Dialog Systems. *AI Magazine*, *37*(4), 7–18. https://doi.org/10.1609/aimag.v37i4.2687.

Weekes, B., & Stauffacher, D. (2018). *Digital Human Security 2020. Human security in the age of AI: Securing and empowering individuals*. ICT4Peace Foundation. https://ict4peace.org/wp-content/uploads/2018/12/Digital-Human-Security-Final-DSmlogos.pdf.

Wilkers, R. (2018, September 20). *Interest is growing in unmanned maritime systems*. Defense Systems. https://defensesystems.com/articles/2018/09/21/unmanned-market-grows.aspx.

Work, R. O. (2017, April 26). *Establishment of an Algorithmic Warfare Cross-Funcion Team (Project Maven). Memorandum of Apr. 26, 2017*. https://dodcio.defense.gov/Portals/0/Documents/Project Maven DSD Memo 20170425.pdf.

CHAPTER 3

AI Talent Competition: Challenges to the United States' Technological Superiority

Ilona Poseliuzhna

Global order is one of the key concepts in the discipline of international relations (IR), and the role of technology in the theory of IR has been increasingly recognized by scholars (McCarthy, 2015). The exact possibility of technology's impact on the international order, however, remains the subject of a debate between instrumentalists, essentialists (including techno-optimists and techno-pessimists), social constructivists (SCOT) and other prominent theorists (e.g. Morgenthau, 1946; Nye, 2010; Rosenau, 2005).

This text intends to contribute to the discussion on the evolution of the international order in response to the development of artificial intelligence (AI), which is more and more often referred to as a disruptive technology (for more on disruptive technologies and AI, see e.g. Girasa 2020). AI plays a role in other processes (Feijóo and Kwon, 2020) that transform the global order such as, inter alia, the growing economic and technological potential of China and multipolar tendencies. These processes present a threat to American domination, which has been relying significantly on technological advantage over America's would-be rivals in 20–21st century. Growing understanding of these processes by American decision makers results in the preparation and implementation of strategies strengthening their key asset – innovation. Meanwhile, revisionist countries (such as China) are aware that technological superiority is one of the main determinants of the US economic and military position. Hence, these states see matching or surpassing US AI technology as an opportunity to improve their standing on the international arena. AI is indeed a promising technology on all levels, but political will, good state of related technologies, adequate investment, and human talent are needed to successfully develop and deploy it. Taking into consideration the political nature of technology development, this article focuses on the first and last factors: political will and human talent. It recognizes that AI talent is an essential resource, both for China (to win the technological race with the USA) and for the United States (to remain the leader of the international technological order).

While in recent years there have been reports measuring the size of the talent pool of both countries and the migratory flows of a highly skilled workforce

 | DOI:10.1163/9789004547261_005

emerged, the results are often lacking in relation to the broader context of IR and professed values. In an attempt to fill this gap, the chapter will first outline the Chinese and US perceptions of the role of AI as well as contemporary shifts in AI talent. It will then show how these trends could threaten the US position and strengthen China. Finally, a brief discussion will be made on the relevant values. Essentially, the question is whether the US should diverge from the democratic values permeating its academic culture (at least relative to that of China, as highlighted by Abungu (2021)) in order to maintain its technological superiority.

1 International Technological Order: USA and China Facing AI

Technology is at the heart of the current developments in IR. Capturing its relationship with key concepts developed in IR such as power, security and global order could be an daunting challenge. However, the concerns shared by eminent scholars should help establish the study of technology as one of the cornerstones of contemporary IR. The relationship between power and technology is not new to the discipline – in fact, there is a whole tradition of engagement with technology throughout the 'classics' of IR. But the digital realm and some of its emerging technologies could become strategic assets (or blatant vulnerabilities) in that sense (Kaltofen et al., 2018). This is particularly evident in the case of innovations with high transformative potential and the fact is already being appreciated by actors on the international stage.

Apart from measurable factors, the mindset of strategists and decision-makers with regards to technology is extremely important: how they think about the technology, how much importance they ascribe to it and how their level of faith in its usefulness drives the technological advances. The recent transformations in IR have resulted in an increased discussion on the social sources of power, which in turn led to the growing recognition that technology is also by its very nature social. In fact, "Technological change has been key to transformations of the international order, in terms of how we conceptualise ourselves as political units as well as shifts in material processes and relational power dynamics" (Kaltofen et al., 2018, 70). From an international politics perspective, recognition of the importance of innovations by decision-makers favors the further formalization of certain ideas. The internalization of beliefs regarding the effectiveness of technology, along with its expected benefits or threats, often results in creation of strategies corresponding to the current needs and challenges. In this regard, it is worth looking at the US approach to the building of international technology order favorable for them.

The US Department of Defense (DOD) refers to the periods of US technological dominance in the world as *offsets*. Here, offsetting means balancing the potential strengths of the opponent (e.g. numerical advantage) with one's own technological advances (Adamsky, 2010). This superiority has become an indicator of US domination in the 20th and 21st centuries. The First Offset included the technology of nuclear weapons and intercontinental ballistic missiles (ICBMs). The Second Offset included the creation of stealth technology and precision-guided munitions. Offset strategies are examples of Cold War thinking: when the Americans did not have to match the Soviet military's man-for-man numerical advantage due to their technological advantage. As this chapter raises the subject of US-Chinese AI competition, it is worth adding that the initial attempts to define a similar way of thinking among the Chinese decision-makers came with President Jiang Zemin's 1993 proclamation of the need to win "local wars under high-tech conditions." This shift in Chinese military doctrine came after the US demonstrated the devastating effectiveness of high-tech weaponry against a less-advanced adversary in the Gulf War (Huang, 2001).

In his 2014 speech, US Secretary of Defense Chuck Hagel emphasized the role of disruptive technologies for security and mentioned the attempts by China and Russia to close the technological gap in order to evolve the international technological order. Hagel alerted his fellow countrymen to the coming of "an era where American dominance" on the seas, in the skies, in space and cyberspace "can no longer be taken for granted" (Hagel, 2014). This speech spurred the development of the so-called Third Offset, based on AI. Its principles were reflected in the 2018 US National Defense Strategy (Mattis, 2018). The key elements of the strategy are autonomous learning systems or machines that can adapt to changing circumstances and human/machine collaborations. The Third Offset is targeting several promising technology areas, including robotics and system autonomy, miniaturization, big data, and advanced manufacturing. In 2018, Congress created a commission to examine how the US could keep pace in AI and related technologies with global competitors. The result was a 756-page blueprint for "defending America in an AI era" and "winning the technology competition" (Schmidt et al., 2021, p. 10–11).

The report addresses emerging national security threats in AI and proposes a strategy for winning the technology competition, focusing on securing talent, promoting, inter alia, innovation and protection of intellectual property rights. National Security Commission on Artificial Intelligence (NSCAI) identified several areas where action is necessary: leadership, talent, hardware, and innovation investment. Of those, NSCAI identified human talent deficit as "the US government's most conspicuous AI deficit and the single greatest inhibitor to buying, building, and fielding AI-enabled technologies for national security purposes" (Schmidt et al., 2021, p. 3). NSCAI, setting goals for preparing for the AI-shaped reality by 2025, points towards various challenges. These challenges are already hindering the ambitious objectives of the Third Offset Strategy in seeking to outmaneuver the advantages made by top adversaries through new technologies. In terms of evolution of the international technological order, NSCAI shares the conclusion that the political will to prioritize AI and associated technologies will allow the US to yet again maintain superiority as this has been the case for decades. The message of the published document to the international community and to the main adversary, China, is clear: the US perceives research, development, and deployment of AI as a competition and is committed to winning.

However, China is no stranger to the ideas of AI competition. Even earlier than the Americans, China's New Generation Artificial Intelligence Development Plan stated that "AI has become a new focus of international competition" (Webster et al., 2017). From the Chinese point of view, AI development is an opportunity to reduce its vulnerable dependence on imports of international technology and to eventually become the world leader in the field by 2030. This is part of a broader revisionist agenda that challenges the current international order and promotes multipolarity (see Clegg, 2009). As mentioned before, Chinese efforts to develop technology, modernize the armed forces, and increase defense spending began already in the 1990s (for more on defense R&D, see Acharya & Arnold, 2019). These efforts came in various forms, most recently – in the form of boosting domestic AI implementation capabilities. Chinese planners "aim to inject AI into every grid in their operational systems at all levels in order to exploit this "strategic front line" technology and create a steep increase in military capability" (Work and Grant, 2019: 14). To this end, institutional efforts have been made: for example, Ministry of National Defense has established two major research organizations focused on AI and unmanned systems under the National University of Defense Technology. The Bases and Talents Program supports the establishment of a series scientific research bases with international competitiveness, and the fostering of top-notch innovative talents and teams. However, the publicly available data on the initiatives within the Program's R&D pillar remains scant. Chinese technological mobilization has yielded many successes, some of which will be discussed in the next section. Particular emphasis is put on the expansion of the talent base and scientific activity.[1]

1 The Chinese have published 3.5 times more AI peer-reviewed articles in recent years. At the same time, the US recorded a smaller increase in publishing – 2.75 times (Zhang et al., 2021, p. 20).

Although offset strategies were by design principally concerned with the military sphere, nowadays the perception of technological advantage is evolving due to the disruptive impact of AI on the economy and civilian life. For years, scholars have been trying to answer the question as to the drivers of choices made by states regarding their technological advancement. Presumably, the actions taken depend both on real (material) events and factors and on the subjective perception of reality (including views on technology itself) by governments and individual decision-makers. If that is the case, then concordant American and Chinese belief in the AI capabilities, combined with mutual perception of each other as competitor, should fuel the AI race between those powers.

2 Closing the Gap

Two decades ago, there was a chasm between China and the US in AI research and talent pool. Today, at least on the surface, the US remains a leader in attracting talent. The leading indexes measuring progress in AI development (e.g. Tortoise, 2021; Shearer et al., 2020) still place the US ahead of China. However, the gap is closing quickly (Castro and McLaughlin, 2021) as China stands a reasonable chance of overtaking the US as the leading center of AI innovation in the coming decade. Growing interest in AI is reflected not only in the strategies of the Communist Party of China, but also in the respective data. Several years ago, China superseded USA in the total number of AI journal publications. *Nature Index* ranks China first and the US second by total number of publications in the field of AI between 2015 and 2019 (*Nature Index 2020 Artificial Intelligence*, 2020). In 2020, for the first time, China surpassed the US in the global share of AI journal citations, with the largest share of conference publications among the major AI powers. The US has by far the biggest talent pull of all countries, but the gap in AI research, development and – particularly – deployment between the US and China is closing (Gagné, 2020). Although the subject of the country's growing R&D capabilities could only be of interest to think tanks dealing with US-Chinese rivalry, there are a few complications that can contribute to massive changes in the existing technological order that currently still remains advantageous to the West. These challenges concern both the geographic dimension (e.g. the distribution of technological capabilities across different regions) and the professed values. More thorough consideration leads to the conclusion that year after year trends indicate a relative decline in advantage in some areas.

Of the many challenges facing the US in the (post)COVID era, reducing the gap between the slowly expanding STEM talent pool and increasing job demand seems to be one of the most obvious. Although American universities offer quite varied educational programs, there are still not enough STEM graduates to fill the gap. For example, some indicators show that less than 17% of computer scientists graduate from American universities each year in relation to the total demand on the labor market (Schmidt et al., 2021: 126). Through the lens of global competitiveness, job demand for technical talent has also been steadily increasing in recent years (Gagné, 2020). Talent shortage becomes visible both at the national level – in sectors of strategic importance for the state – and at the local level, in the growing digital transformation pressure experienced by entrepreneurs.

Between 2000 and 2014, the growth in number of science and engineering bachelor's degrees awarded in China exceeded 360%. Meanwhile, the US experienced more moderate growth (54%) over the same period (National Science Board, 2018). Such difference is easily explained by the rapid economic development in China and the corresponding growth in demand for a skilled workforce in this field, but the increase in scientific pursuits still should be considered above average. STEM graduates are the key human resource of digital transformation, but it is the rate of research output that ultimately shows who is leading the race in research. Is it still the West, then? Chinese academic output has grown almost twice as fast as the world's annual average for the last 10 years. At the same time, output of the US and European Union (EU) has grown at less than half of the world's annual growth rate (National Science Board, 2019a), slow enough for some experts to start raising their concerns. "For the first time in our lifetime, the United States risks losing the competition for talent on the scientific frontiers," (Schmidt et al., 2021, p. 173) – this NSCAI conclusion sounded the alarm for the traditional American technological supremacy, in stark contrast with the previous offsets.

Still, R&D is not enough to attain superiority in any field. Effective application is no less essential. At the current level of advancement in narrow artificial intelligence, countries need skilled engineers focused on implementation in order to build cohesive AI ecosystems. While China is not a leader in conducting innovative research, the aforementioned growth of STEM graduates of Chinese universities should result in an influx of vast numbers of implementers into the industry. They will serve as AI engineers, the lifeblood of the digital economy: training models on a daily basis, finding new applications in response to emerging needs, analyzing and cleaning data. This valuable human resource will produce little to no R&D but will be able to implement solutions and fix local errors at all levels of day-to-day economy. Fundamentally, these trends are already visible: China is the country with the largest share of implementers (Gagné, 2020). Formerly, AI as an industry was concentrated around high-level experts who were capable of making breakthroughs in fundamental research and of managing new techniques. Over time, AI specialists' skillset and professional roles evolve more toward applied sciences. Chinese STEM education plays into this trend, to the advantage of the country's economy and international position.

It is worth noting in this regard that the US government faces the challenge of talent competition primarily from the US private sector. NSCAI points out that the main factor discouraging young professionals from working in the public sector is the belief that it is difficult to perform meaningful work focused on their field of expertise in government (Schmidt et al., 2021: 122). However, it is impossible to address the problem without mentioning other obstacles, such as the significant pay gap between BigTech and the public sector. IT specialists also expect greater awareness from political leaders about the impact of new technologies on social and political reality, having better access to tools, data sets, and infrastructure to train their models on as well as the possibility of planning their careers exclusively in their chosen field. Having these preferences defined gives the opportunity to adopt more effective tools to prevent further outflow of AI talent from industries and services controlled by government. Some expectations, however have a less tangible dimension – AI talent wants to be appreciated by decision-makers and values working in an open and conducive environment.

The unprecedented brain drain of AI professors from university to the industry (Gofman and Jin, 2020; Procaccia, 2019) threatens to inhibit fundamental research capacity on AI in America, of which American universities are the primary source. In 2020, the highest proportion of peer-reviewed AI papers came from academic institutions (Zhang et al., 2021, p. 10). Although the share of papers from large technology companies is increasing (Ahmed and Wahed, 2020), business is more interested in implementing the existing narrow AI (NAI) solutions than in conducting costly, long-term fundamental research. The private sector is led by a certain logic, according to which there is no guarantee that this research will be successful and therefore profitable (Eynon and Young, 2021). While understandable, this logic could discourage innovativeness, which was an important American advantage in building a favorable technological order. What is more, teaching base will suffer as well due to the brain drain, further limiting the quality of basic research produced by academia. Over the past decade, industry has begun to play a greater role in attracting AI talent: 65% of North American PhD graduates in AI chose to work in industry – up from 44.4% in 2010 (Zhang et al., 2021: 4). The statistics of this "domestic" brain drain in North America are rather alarming, as it is estimated that the share of new AI PhDs entering academia dropped to 23.9% in 2019, compared to 42.1% in 2010 (Zhang et al., 2021: 12). The overlapping challenges of talent pool composition, quantity of research output and the outflow of talent from US universities create a complex picture of weaknesses and threats to US innovation in the era of AI competition. The growing importance of STEM skills in the workforce has not been matched by greater participation of American students in STEM, neither in schools, nor during tertiary education. Inconsistent attempts to develop more engaging school curricula and pedagogy are one of the possible reasons for the failure to attract students to STEM (see Barkatsas et al., 2018) as well as to retain PhDs at the intra-state level.

The US still has a chance to compensate its domestic shortcomings by recruiting from international talent and enabling expats to realize their full potential (Allison and Schmidt 2020). Moreover, articles of great scientific interest are generally produced by international researchers and often as part of cross-border collaboration. Openness in science matters to a nation's scientific influence (Wagner and Jonkers, 2017) and this openness for many years was an undeniable advantage of the American state, contributing to its soft power and boosting its international position. In this context, the recent anti-immigration narratives and political actions of the President Donald Trump's administration should be assessed critically from the perspective of stimulating the innovation of American science. Escalating immigration obstacles for skilled workers and students only reinforce the negative trends outlined earlier. The rationale behind these restrictions was, among other things, to prevent illegal transfers of American technology (for more about the real possibilities of copying American technology and the complexity thereof, see Gilli and Gilli 2018). However, there are reasonable doubts as to whether such simplistic solutions correspond to the complexity of the problem. The one thing that can be confirmed is that they negatively affect the net migration rates of highly qualified specialists to America. In fact, these actions, presented as a boost for national security, might even prove counterproductive in this regard. They contradict the key conclusion on STEM migration flows and American technological superiority: the US gains more than it loses from attracting Chinese talent – and those gains come at China's expense. Data indicates that studying in the US positively influences the rate of graduates staying in the country – only about 9% of them decide to work in other countries post-graduation (Zhang et al., 2021: 12). Furthermore, it is often forgotten that the most sensitive research from a national security perspective is actually carried out by senior researchers and engineers and that students do not participate in it (Zwetsloot and Arnold, 2021). At the same time, China has reformed immigration rules to attract more foreign tech talent. Actively removing immigration incentives for highly qualified professionals is therefore detrimental US competitiveness.

American competitive advantage in the field of AI has been built, among other things, on attracting international students and researchers to work for US institutions (Banerjee and Sheehan, 2020). This could be one of the reasons why it is not in the US national interest to limit the numbers of Chinese entering American universities.

The data shows that, under new restrictions, foreign students and scholars felt less safe and less welcome than the previous year surveyed (NAFSA, 2020). While the preceding concerns mainly challenges in the areas of education and research, domestic AI workforce shortages only exacerbates the growing gap and requires action. International students' attraction and retention historically been a core strength of the US academic culture, and now these people are at the forefront of digital transformation. This is confirmed by the fact that more than half of the AI workforce in the US was born abroad, as were around two-thirds of current graduate students in computer science and electrical engineering. Meanwhile, the domestic pool of AI graduate students has remained flat since the early 1990s (Zwetsloot et al., 2019: 17). The US should certainly focus on strengthening domestic talent growth in order to maintain its technological leadership position and to prevent irreversible vulnerability in this area. Developing talent takes years, and the US do not necessarily have that much time available. Although many claims and media headlines regarding Chinese technological supremacy can be viewed as alarmist, time is indeed the key resource that allows China to position itself better in the AI competition. For this reason, the current policy of forfeiting the key advantage of US in acquiring international talent for the sake of uncertain security gains could likely prove contrary to the actual the US security interests.

Simultaneously, the effectiveness of human resources in technological competition is also dependent on the quantity and quality of big data available for machine learning. There is a general agreement among Western researchers that Chinese long-term advantage lies in its 1.4 billion population. Population is a resource for digital economy – it determines the size and variety of the data pool which can be used to train algorithms. The large datasets are an indispensable part of AI development, because training software with machine learning requires data. Companies with access to large datasets have a competitive advantage over those without it. Chinese AI sector is thus powered by the world's largest domestic market and human resources. Less commitment to the value of privacy is often perceived as an advantage in this competition (Allison and Schmidt, 2020: 12). On the surface, the abundance of data generated by a large population and a more developed Internet of Things architecture give Chinese a great advantage compared to, for example, aspiring European countries with stricter approach to privacy. However, the locality of the data is also important and "the amount of training datasets required depends on the function that a particular AI tool is designed to perform, and some AI tools require a reduced scale of data" (Abungu, 2021: 3; see also Loukissas, 2019).

One of the most recent proposals to solve the big data issue in favor of the US comes from a group of scientists from US research universities and national laboratories. The authors note that the US research and education communities lack access to important elements of new computational fabric, which consist of three elements: "emergence of public cloud utilities as a new computing platform; the ability to extract information from enormous quantities of data via machine learning; and the emergence of computational simulation as a research method on par with experimental science" (Foster et al., 2021). In their white paper, they suggest creating and implementing a National Discovery Cloud (NDC) to accelerate innovation in digital technologies as well as increase the competitiveness of scientists who previously had no access to these capabilities.

When focusing on evaluating the potential of datasets, American commentators often seem to overlook these two factors. It must be admitted, however, that beliefs about the growing Chinese possibilities triggered a response in the US, leading the 117th United States Congress to emerge as the most AI-centric session in history, with the introduction of 130 AI bills. Increasing awareness and the very pivot towards innovation and the search for comprehensive solutions should boost American chances on maintaining the current technological order. Although, in terms of population size, the US is not on par with China, and the progress in adopting AI in recent years does not match the real capabilities of the Americans, a close eye should be kept on one of the most decisive American advantages so far – talent and democratic culture.

3 A Matter of Values

The argument thus far has primarily focused on the gap between the slowly expanding STEM talent pool and quickly growing job demand, as well as to disruptive processes that only magnify this divergence in the US. To reiterate, the relevant factors include:

- the brain drain of AI scientists from academia to industry, which results in less involvement in basic research (which is one of the factors behind the American innovations in the 20th and 21st centuries);
- immigration policy measures that disfavor international students (who are the basis of human resources in AI);

- the determination of Chinese authorities to gain an advantage in AI using comprehensive measures of talent financing, such as subsidies for technology companies and academic institutions that engage in cutting-edge AI research. China's push for global leadership in AI is further reinforced by factors such as quick organization and persistence in achieving the objectives.

The political nature of AI is expressed in setting standards and promoting values, cooperating in innovations, coordinating policies and approach to development with the international community. Political drivers influence both technical and ecosystem factors (e.g. R&D, AI talent pool, STEM education, investment). In its political nature, the discussion on AI takes the form of a discussion on the values underlying the actions taken. Recently, in the public debate in the West, one encounters with the view that that the US democratic culture stands in the way of decisive action in the development and deployment of AI in some areas (Faggella, 2019; Allison and Schmidt 2020)[2]. The question is, therefore, raised as to whether the US should diverge from the democratic values permeating its academic culture in order to maintain its technological superiority.

To answer this question, it must first be understood where this belief comes from. In this sense, values are considered a normative factor, but as such they give impetus to technical factors, the shaping of the STEM and AI education frameworks, political decisions on migration, etc. Three key factors determining the current state of AI affairs in China need to be pointed out: the unique nature of AI research; a vibrant China's market suppplying the AI sector with funding, workers and data; and lack of strict regulations allowing for wider use of ANI's achievements (Li et al., 2021). The entire entrepreneurial ecosystem, the determination of the authorities and weak privacy regulations have enabled China to catch up with the forerunner the USA, at least in some areas. However, it is worth remembering that this success has been enabled by China's access to global technology research and markets as well as reliance on open source technologies, often developed by Americans or international research groups with American participation. China has successfully reaped

2 Beliefs about a better alignment of Chinese culture and innovation system with the implementation of ANI achievements were expressed in an influential bestseller *AI superpowers: China, Silicon Valley, and the new world order* (Lee, 2018). The arguments put forward by the author were quickly disseminated by the mainstream media. They intensified the perpetuation of the vision that the US may lose in the AI race because of its democratic culture, commitment to regulation and the pursuit of respect for human rights.

significant benefits of the costly basic research conducted over decades in the West, which enabled its own AI sector to focus primarily on the application side. This means that the openness of global AI science has indirectly helped China in reducing the knowledge gap.

There is another widely held belief that higher American threshold for individual privacy rights leaves the US at a disadvantage in collecting datasets and that China has many more possibilities in this field, supported by a large population. However, it has already been highlighted in this chapter that the popular argument of Chinese technological acceleration being driven by the sheer size of its population is not conclusive in AI competition. The issue is more nuanced and there are possible technological solutions to help overcome this obstacle. There is a risk that AI competition will put to the test the US approach to intellectual property, free market principles, individual liberty, and limited government. Although there is no one consistent concept of American individualism, its core values seem to be dignity, autonomy and privacy. Hence, in the US, a significant amount of attention is paid to ethical aspects and values in the development and implementation of AI, which is noticeable in documents issued by state authorities and the activity of civil society. China, on the other hand, focuses on the dynamics of development and surpassing the US whilst ignoring ethical concerns.

However, the Chinese approach also has potential weaknesses, which could allow countries with democratic values to stay in the game. The openness of science does not create incentives for business to invest in basic AI technologies. AI patents are owned mainly by the state-sponsored or government-owned universities, which by itself is not a barrier to innovation (Li et al., 2021). Still, cooperation between companies and universities in China is relatively weak and the transfer of knowledge and technology is insufficient to achieve synergies.

Moreover, although the Communist Party of China is creating new ways to educate and retain top-tier AI talent (China Institute for Science and Technology Policy, 2018), a high proportion of STEM graduates and researchers from Chinese universities keep leaving China. Moreover, empirical studies have revealed that the Chinese research culture hampers talent retention in contrast to the cultural environment found in democratic countries. (Ding, 2018; see also Abungu 2021). While tracking the mobility of AI workforce is challenging, some estimates indicate that up to three-quarters of Chinese AI researchers have chosen to work outside of their homeland. This situation could partially result from resource allocation, most of which the government has spent on expanding the talent base rather than on creating incentives for retaining such talent within the country (Dantong, 2019).

With these issues in mind, one should recall that the US still remains home to the highest number of top-tier AI researchers and an attractive destination for top foreign specialists. As for today, most fundamental breakthroughs in AI R&D have been made in the USA. It should be noted that, if in the long run the US will manage to make further breakthroughs in this field, then many of today's warnings about the evolution of the global technological order will be rendered obsolete. After all, "In AI, brain power matters more than computing power" (Allison and Schmidt 2020: 10). For this reason, democratic values and the individualistic principle of self-development (e.g. AI talent) should not be overlooked as possible sources of long-term American advantage, with broader implications for the international order.

All this leads to the conclusion that the democratic values of the US do not necessarily stand in the way of maintaining their technological superiority but might even provide a basis for effective policies. Moreover, the commitment to democracy may strengthen ties with America's long-standing allies and pave ways to cooperation that could be based on the adaptation of emerging technologies to common values. In order to maintain its leadership position, the US should play to its strengths: openness and a long-term focus in science, attracting foreign talent, and using multiple forums to engage with international partners. The democratic culture of its academia gives the US an opportunity to press home the advantages in key areas where the Chinese model, as of yet, shows its shortcomings.

Acknowledgement

The publication was funded by the Priority Research Area Society of the Future under the program "Excellence Initiative – Research University" at the Jagiellonian University in Krakow.

References

Abungu, C. Y. (2021). Democratic Culture and the Development of Artificial Intelligence in the USA and China. *The Chinese Journal of Comparative Law*, 1–28. https://doi.org/10.1093/cjcl/cxaa032.

Acharya, A., and Arnold, Z. (2019, December). *Chinese Public AI R&D Spending: Provisional Findings*. Center for Security and Emerging Technology. https://cset.georgetown.edu/wp-content/uploads/Chinese-Public-AI-RD-Spending-Provisional-Findings-1.pdf.

Adamsky, D. (2010). *The culture of military innovation: The impact of cultural factors on the revolution in military affairs in Russia, the US, and Israel.* Stanford: Stanford University Press.

Ahmed, N., and Wahed, M. (2020). The De-democratization of AI: Deep Learning and the Compute Divide in Artificial Intelligence Research. *arXiv preprint arXiv:2010.15581.*

Allison, G., and Schmidt, E. (2020, August). *Is China Beating the US to AI Supremacy?* Harvard Kennedy School Belfer Center for Science and International Affairs. https://www.belfercenter.org/sites/default/files/2020-08/AISupremacy.pdf.

Banerjee, I., and Sheehan, M. (2020, June 9). *America's Got AI Talent: US' Big Lead in AI Research Is Built on Importing Researchers.* MacroPolo. https://macropolo.org/americas-got-ai-talent-us-big-lead-in-ai-research-is-built-on-importing-researchers/?rp=m.

Barkatsas, T., Carr, N., and Cooper, G. (2018). *STEM education: An emerging field of inquiry.* BRILL.

Castro, D., and McLaughlin, M. (2021, January). *Who is winning the AI race: China, the EU, or the United States?* Center for Dara Innovation. https://www2.datainnovation.org/2021-china-eu-us-ai.pdf.

China Institute for Science and Technology Policy. (2018, July). *China AI Development Report 2018.* http://www.sppm.tsinghua.edu.cn/eWebEditor/UploadFile/China_AI_development_report_2018.pdf.

Clegg, J. (2009). *China's Global Strategy: Towards a Multipolar World.* London; New York: Pluto Press.

Dantong, J. (2019, July 31). *China's AI Talent Base Is Growing, and then Leaving.* MacroPolo. https://macropolo.org/chinas-ai-talent-base-is-growing-and-then-leaving/?rp=e.

Ding, J. (2018). Deciphering China's AI dream. *Future of Humanity Institute Technical Report.*

Eynon, R., and Young, E. (2021). Methodology, legend, and rhetoric: The constructions of AI by academia, industry, and policy groups for lifelong learning. *Science, Technology, & Human Values, 46*(1), 166–191.

Faggella, D. (2019, November 9). *The USA-China AI Race – 7 Weaknesses of the West.* Emerj – Artificial Intelligence Research and Insight. https://emerj.com/ai-power/the-usa-china-ai-race/.

Feijóo, C., and Kwon, Y. (2020). AI impacts on economy and society: Latest developments, open issues and new policy measures. *Telecommunications Policy, 44*(6).

Foster, I. et al. (2021, May). *A National Discovery Cloud: Preparing the US for Global Competitiveness in the New Era of 21st Century Digital Transformation.* The Computing Community Consortium – CCC. https://cra.org/ccc/wp-content/uploads/sites/2/2021/04/CCC-Whitepaper-National-Discovery-Cloud-2021.pdf.

Gagné, J. F. (2020, October 30). *Global AI Talent Report 2020*. Jfgagne. Retrieved May 15, 2021, from https://jfgagne.ai/global-ai-talent-report-2020/.

Gilli, A., and Gilli, M. (2018). Why China has not caught up yet: Military-technological superiority and the limits of imitation, reverse engineering, and cyber espionage. *International Security*, *43*(3), 141–189.

Girasa, R. (2020). *Artificial Intelligence as a Disruptive Technology: Economic Transformation and Government Regulation*. Palgrave Macmillan.

Gofman, M., and Jin, Z. (2020, October 26). Artificial Intelligence, Education, and Entrepreneurship. Available at SSRN: https://ssrn.com/abstract=3449440.

Hagel, C. (2014, September 3). *"Defense Innovation Days" Opening Keynote (Southeastern New England Defense Industry Alliance)*. US Department of Defense. https://www.defense.gov/Newsroom/Speeches/Speech/Article/605602/.

Huang, A. C. C. (2001). Transformation and refinement of Chinese military doctrine: Reflection and critique on the PLA's view. *Seeking Truth From Facts: A Retrospective on Chinese Military Studies in the Post-Mao Era*, 131–140. http://www.rand.org/content/dam/rand/pubs/conf_proceedings/CF160/CF160.ch6.pdf.

Kaltofen, C., Carr, M., and Acuto, M. (Eds.). (2018). *Technologies of International Relations: Continuity and Change*. Springer.

Lee, K. F. (2018). *AI superpowers: China, Silicon Valley, and the new world order*. Houghton Mifflin.

Li, D., Tong, T. W., and Xiao, Y. (2021, February 18). *Is China Emerging as the Global Leader in AI?* Harvard Business Review. https://hbr.org/2021/02/is-china-emerging-as-the-global-leader-in-ai.

Loukissas, Y. A. (2019). *All data are local: Thinking critically in a data-driven society*. MIT press.

Mattis, J. (2018). *Summary of the 2018 national defense strategy of the United States of America*. Department of Defense, Washington. https://apps.dtic.mil/sti/pdfs/AD1045785.pdf (September 17, 2020).

McCarthy, D. R. (2015). *Power, information technology, and international relations theory: The power and politics of US Foreign policy and internet*. Palgrave Macmillan.

Morgenthau, H. (1946). *Scientific Man vs. Power Politics*. The Chicago University Press.

NAFSA: Association of International Educators. (2020, March). *Losing Talent 2020: An Economic and Foreign Policy Risk America Can't Ignore*. https://www.nafsa.org/sites/default/files/media/document/nafsa-losing-talent.pdf.

National Science Board. (2018). *The Rise of China in Science and Engineering*. https://www.nsf.gov/nsb/sei/one-pagers/China-2018.pdf.

National Science Board. (2019a, December). *Publications Output: US Trends and International Comparisons* (NSB-2020-6). https://ncses.nsf.gov/pubs/nsb20206.

Nye, J. S. (2010). *Cyberpower*. Belfer Centre for Science and International Affairs, Harvard Kennedy School. https://www.belfercenter.org/sites/default/files/legacy/files/cyber-power.pdf.

Procaccia, A. (2019, January 7). *Tech Giants, Gorging on AI Professors Is Bad for You*. Bloomberg. https://www.bloomberg.com/opinion/articles/2019-01-07/tech-giants-gorging-on-ai-professors-is-bad-for-you.

Rosenau, J. N. (2005). Illusions of Power and Empire. *History and Theory, 44*(4), 73–87.

Schmidt, E., et al. (2021). *National Security Commission on Artificial Intelligence (AI)*. National Security Commission on Artificial Intellegence.

Shearer, E., et al. (2020, September). *Government AI Readiness Index 2020*. Oxford Insights. https://www.oxfordinsights.com/government-ai-readiness-index-2020.

Tortoise. (2021). *The Global AI Index – Global AI Summit*. https://www.theglobalaisummit.com/FINAL-Spotlighting-the-g20-Nations-Report.pdf.

Wagner, C. S., and Jonkers, K. (2017). Open countries have strong science. *Nature News, 550* (7674), 32–33.

Webster, G. et al. (2017, August 1). *Full Translation: China's "New Generation Artificial Intelligence Development Plan."* New America. https://www.newamerica.org/cybersecurity-initiative/digichina/blog/full-translation-chinas-new-generation-artificial-intelligence-development-plan-2017/.

Work, R. O., and Grant, G. (2019). Beating the Americans at their Own Game: An Offset Strategy with Chinese Characteristics. *Center for New American Security, June*, 24.

Zhang, D., et al. (2021). The AI Index 2021 Annual Report, AI Index Steering Committee, Human-Centered AI Institute, Stanford University, Stanford, CA.

Zwetsloot, R., and Arnold, Z. (2021, April 23). *Chinese Students Are Not a Fifth Column*. Foreign Affairs. https://www.foreignaffairs.com/articles/united-states/2021-04-23/chinese-students-are-not-fifth-column.

Zwetsloot, R. et al. (2019, December). Keeping Top AI Talent in the United States. *Center for Security and Emerging Technology*.

CHAPTER 4

Third-Party Certification and Artificial Intelligence

Some Considerations in Light of the European Commission Proposal for a Regulation on Artificial Intelligence

Jan De Bruyne

1 Introduction

Artificial intelligence (AI) is becoming increasingly more prevalent in our daily lives. Just think of the societal impact of ChatGPT. AI can be of benefit to a wide range of sectors such as healthcare, energy consumption, climate change or financial risk management. AI can also help to detect cybersecurity threats and fraud as well as enable law enforcement authorities to fight crime more efficiently (European Commission, 2019; De Bruyne and Bouteca, 2021). AI systems are more accurate and efficient than humans because they are faster and can better process information (Tzafestas, 2015, 147). They can perform many tasks 'better' than their human counterparts – although it also depends on how this term is eventually defined. An example is playing chess or detecting diseases (Mannens, 2021).

Companies from various economic sectors already rely on AI to decrease costs, generate revenue, enhance product quality and improve competitiveness (Ivanov, 2017, 283–285). AI systems and robots can also have advantages for the specific sector in which they are to be used. Traffic, for example, is expected to become safer with autonomous vehicles. The number of accidents may decrease as a result of a computer replacing the human driver and, therefore, eliminating any human error related to accidents (De Bruyne and Tanghe, 2017; Brijs, 2021, 127–145). At the same time, however, the introduction of AI and robots raises many challenges. These will only become more acute in light of the predicted explosive growth of the robotics industry over the next decade (Calo, 2016, 3). Indeed, AI has implications for various facets of our society (Organisation for Economic Co-operation and Development, 2019) resulting in multiple legal challenges (Barfield and Pagallo, 2018; Abbott, 2020; Lavy and Hervey, 2020; De Bruyne and Vanleenhove, 2022; Custers and Fosch Villaronga, 2022) and ethical issues (Dignum, 2019; Dubber, Pasquale and Das, 2020; Coeckelbergh, 2020).

 | DOI:10.1163/9789004547261_006

Against this background, the European Union (EU) has already taken several initiatives to regulate artificial intelligence and move towards trustworthy and human-centric AI. After a brief overview and discussion of some of these initiatives, I will focus on the use of third-party certification as a regulatory tool to achieve trustworthy and human-centric AI. Third-party certification has been extensively relied upon in many sectors (see for an overview De Bruyne, 2019, 7–17). As rightly mentioned by Möslein and Zicari, however, 'the design of certification schemes for AI technologies [...] has hardly been investigated at this point' (Möslein and Zicari, 2021, 357). To that end, I will discuss the concept of third-party certification and subsequently examine how tort law can be relied upon as a mechanism to increase the accuracy and reliability of certificates. I will conclude this chapter by summarising the most important findings.

2 Towards a Proposal for a Regulation of AI in the European Union

In its Communication of April 2018, the European Commission (EC) puts forward a European approach to AI based on three pillars: (1) being ahead of technological developments and encouraging the uptake of AI by the public and private sectors; (2) preparing for socio-economic changes resulting from AI and (3) ensuring an appropriate ethical and legal framework (European Commission, 2018, 2). The EC subsequently established the High-Level Expert Group on Artificial Intelligence (AI HLEG) to support the implementation of this strategy. The AI HLEG issued several documents among which were the Ethics Guidelines for Trustworthy Artificial Intelligence in April 2019 (AI HLEG, 2019). According to the Guidelines, trustworthy AI should be ethical, lawful and robust from both technical and social perspectives. The Guidelines list seven key (ethical) requirements that AI systems should meet to be trustworthy: (1) human agency and oversight; (2) technical robustness and safety; (3) privacy and data governance; (4) transparency; (5) diversity, non-discrimination and fairness; (6) societal and environmental well-being; and (7) accountability. The requirements are applicable to the different stakeholders partaking in AI systems' life cycle (e.g. developers, deployers, end-users, and the broader society). After a piloting process, the AI HLEG presented its final Assessment List for Trustworthy Artificial Intelligence (ALTAI) on 17 July 2020. Through ALTAI, the seven principles are translated into an accessible and dynamic checklist that guides developers and deployers of AI systems in implementing such principles in practice (AI HLEG, 2020).

The EC issued its White Paper on AI in February 2020 paving the way to a proper proposal of a legal framework on AI (European Commission, 2020).

The White Paper contains policy options to enable a trustworthy and secure development of AI in the EU, thereby taking into account the values and rights of EU citizens. The White Paper consists of two main building blocks, namely an 'ecosystem of trust' and an 'ecosystem of excellence'. Creating an ecosystem of trust will give citizens the confidence to take up AI systems and provide companies and public organisations the legal certainty to innovate by using AI. The EC stresses that any regulatory intervention should be targeted and proportionate. To that end, it adopts a risk-based approach implying that it aims to regulate high-risk AI systems. Systems that are not considered high-risk should only be covered by more general legislation, for example on data protection, consumer protection, and product safety/liability. These rules may, however, need some targeted modifications to effectively address the risks created by AI. The Commission also proposes several recommendations to establish an ecosystem of excellence, for instance by focusing on the required skills and securing access to data.

Based on the White Paper and the outcome of the public consultation process, the EC issued a Proposal for a Regulation laying down harmonised rules on AI, also know as the AI Act (AIA) Proposal (European Commission, 2021; Veale and Zuiderveen Borgesius, 2021; Ebers et al., 2021). The Proposal contains an open definition of AI (Art. 3(1)). AI is defined as software that is developed with one or more of the techniques and approaches listed in Annex I (e.g. machine learning) and can, for a given set of human-defined objectives, generate outputs such as content, predictions, recommendations, or decisions influencing the environments they interact with. The AI Act Proposal also includes measures to support innovation such as the establishment of regulatory sandboxing schemes (Art. 53–55). The Proposal establishes a European Artificial Intelligence Board that will supervise the application of the Regulation across the EU, gather best practices and advise EU institutions on questions regarding AI (Art. 56–58). The Proposal imposes obligations upon several parties such as providers and users of high-risk AI systems (Art. 16–29). Those obligations will be important when it comes to establishing the potential liability of such parties (De Bruyne, Van Gool and Gils, 2021, 407–408). Member States also have to lay down effective, proportionate and dissuasive penalties, including administrative fines, when AI systems that are put on the market do not respect the requirements of the Regulation (Art. 71–72).

More importantly, the Proposal relies on a risk-based approach. Art. 5 prohibits certain AI systems, such as those used by public authorities for 'social scoring'. Art. 6 in turn refers to high-risk AI systems, which is the case when two conditions are fulfilled: (a) the AI system is intended to be used as a safety component of a product or is itself a product covered by the EU harmonisation legislation listed in Annex II and (b) the product whose safety component is the AI system or the AI system itself as a product is required to undergo a third-party conformity assessment with a view to the placing on the market or putting into service of that product pursuant to the EU harmonisation legislation listed in Annex II. This Annex contains a list of product safety and related market surveillance legislation including the Medical Device Regulation or the Machinery Directive. Annex III to the Proposal also refers to specific AI systems that are identified as high-risk (e.g. those used for recruitment or in education). High-risk AI systems need to comply with the additional requirements included in Chapter 2 of the Proposal. These *inter alia* include the establishment of a risk management system, the need to ensure human oversight and the requirement of record keeping. Providers of high-risk AI systems must ensure that these undergo the relevant conformity assessment procedures prior to their placing on the market or putting them into service (Art. 16). This will in some cases require the involvement of a notified body, which will be further discussed.

The Proposal also imposes specific transparency obligations for certain AI systems (Art. 52). Transparency obligations will apply for AI systems that (i) interact with humans, (ii) are used to detect emotions or determine association with (social) categories based on biometric data or (iii) generate or manipulate content ('deep fakes'). For instance, when persons interact with an AI system or their emotions or characteristics are recognised through automated means, they must be informed of that circumstance. Finally, the proposal refers to AI systems with a minimal or no risk, which are permitted without restriction. Such AI systems can be developed and used subject to the existing legislation without additional legal obligations (e.g. consumer protection, data protection, etc.). The vast majority of AI systems currently used in the EU fall into this category. Pursuant to Art. 69, providers of those systems can voluntarily adhere to codes of conduct (Hawes, Hatzel and Holder, 2021; Buri and von Bothmer, 2021).

The AI Act proposal by the European Commission is currently being addressed by and discussed in the competent committees in the European Parliament as well as by the Council of the EU. In this regard, the Council adopted its common position in December 2022. It contains several amendments. AI, for instance, is defined as a system that is designed to operate with 'elements of autonomy' and that, based on machine and/or human-provided data and inputs, infers how to achieve a given set of objectives using machine learning and/or logic-and knowledge based approaches, and produces system-generated outputs such as content (generative AI systems), predictions, recommendations or decisions, influencing the environments with which the AI system interacts (Art. 3(1)). It also broadens the scope of prohibited AI systems (Art. 5) and includes an additional horizontal layer for stand-alone AI systems. Such systems will be considered high-risk unless the output of the system is 'purely accessory' with regard to the relevant action or decision to be taken and is not therefore likely to lead to a significant risk to the health, safety or fundamental rights (Art. 6.3). The text also contains provisions on General Purpose AI (GPAI) in Art. 4 and adapts the requirements for high-risk AI systems (Art. 8–15).

3 The Reliance on Third-Party Certifiers

It has already been mentioned that providers of high-risk AI systems must ensure that their systems undergo the relevant conformity assessment procedure prior to their placing on the market or putting into service (Art. 16). This will allow providers to demonstrate that their systems comply with the mandatory requirements for trustworthy AI (e.g. data quality, accuracy and robustness). The AI Act Proposal contains several provisions on how this conformity assessment has to be done (cf. Art. 33, Art. 43 and Art. 44).

For high-risk AI systems referred to in points 2 to 8 of Annex III of the Proposal AI Act (e.g. management and operation of critical infrastructure or education and vocational training), providers have to follow the conformity assessment procedure based on internal control as referred to in Annex VI. This procedure does not involve a notified body. Providers have to self-assess that their quality management system, system-specific technical documentation and post-market monitoring plan follow either the essential requirements or a relevant harmonised standard/common specification. The Proposal for an AI Act gives AI providers an unduly broad margin of discretion regarding its application and enforcement, especially regarding: whether the used software is an AI system; whether the system may likely cause harm; and how to comply with the mandatory requirements for of Title III, Chapter 2 of the Proposal for an AI Act. However, a mere internal control by providers of stand-alone AI systems may be problematic. According to different scholars, it is questionable whether reliance on self-certification provides a meaningful legal assurance that the requirements to obtain a CE mark for high-risk AI systems are properly met (e.g. Smuha et al., 2021, 39). The European Commission should thus explore whether at least certain other high-risk AI systems need to be subject to an independent ex ante control (Ebers, 2021, 595).

This conformity assessment will in some cases require the involvement of independent third-party certifiers, referred to as a notified body in the

Proposal. High-risk applications within the area of 'biometric identification and categorisation of natural persons', for instance, will need to undergo a conformity assessment by a notified body (Art. 43). Once harmonised standards or common specifications are adopted covering those systems exist, however, only self-assessment is needed. Considering that the European Commission envisages that harmonised standards will exist before the application of the Proposal AI Act, specific 'AI notified bodies' as required under the Act may indeed never be required, even not for biometric systems (Veale and Zuiderveen Borgesius 2021, 106). AI systems being safety components of products covered by sectorial Union legislation will always be deemed high-risk when subject to third-party conformity assessment under that sectorial legislation.

In addition to standardisation (Ebers, 2022, 321–344), certification of AI systems will thus become important once the AI Act Proposal has been adopted. Especially third-party certification by an independent body will become crucial. The assessment conducted by an independent notified body is intended to be more rigorous and impartial than self-assessment. It is, therefore, especially appropriate for high-risk AI systems (De Bruyne, 2019, 4–7; Hawes, Hatzel and Holder, 2021; Ebers, 2021 595). In the following paragraphs, I will first focus on the concept of third-party certification. Once this has been done, I will examine how tort liability can induce third-party certifiers to issue more accurate[1] and reliable[2] certificates, which in turn may increase trust in AI systems and hence the societal acceptance of these technologies. The Proposal for a Regulation on AI will then be briefly evaluated.

3.1 *Third-Party Certification as a Regulatory Mechanism*

For the purpose of this discussion, certifiers are entities that provide certification services. They attest that a certified product, service, information, or person (further referred to as 'item') possesses certain qualifications or meets safety, quality, or technical standards (Garner, 2009, 220–221; Greenberg, 2010, 304–305). The certification process can take different forms. First-party certification means that the certification is done by the manufacturers of the products or providers of services/information themselves. Second-party is performed by the party purchasing the products or relying on the services/ information to ensure their compliance with the agreed contractual requirements

1 Accuracy is the quality or state of being correct or precise or the degree to which the result of a measurement, calculation or specification conforms to the correct value or standard. See the definition of 'accuracy' in the online Oxford Dictionary.

2 Reliable is something that can be trusted or believed because it works or behaves well in the way one expects. See the definition of 'reliable' in the online Cambridge Dictionary.

and technical specifications. Third-party certification is performed by organisations that are independent vis-à-vis the entity manufacturing the products, offering the services, or providing information (further referred to as the 'requesting entity'). Third-party certifiers provide their services at the request of clients (De Bruyne, 2019, 3–6). To that end, they enter into a certification agreement with the requesting entities. This contract contains the obligations of both parties during the certification process (Verbruggen, 2013, 229–230). The certificate they issue is the performance under the certification contract. Most certified items bear the certifier's mark to help consumers or other buyers making decisions (De Bruyne, 2019, 4).

Although the issuance of a certificate is the performance under the certification agreement, the information included in this attestation can and will also be used by persons with whom certifiers do not have any contractual relationship (further referred to as 'third parties'). Due to information asymmetry, third parties do not always have all the necessary information on the quality of a particular item (cf. Akerlof's 'market for lemons' problem). Third parties and requesting entities do not possess equal amounts of information required to make an informed decision regarding a transaction. Whereas the requesting entity knows the true value of the item or at least knows whether it is above or below average in quality, a third party, by contrast, does not always have this knowledge as it is not involved in the production of an item. As a consequence, requesting entities might have an incentive to market items of less than average market quality. A third party is not always able to effectively identify 'good' items from those that are below average in quality, which is referred to as the adverse selection problem (Akerlof, 1970, 488; Lofgren, Persson and Weibull, 2002, 195; De Bruyne, 2019, 4). A certifier moderates informational asymmetries that distort or prevent efficient transactions by providing the public with information it would otherwise not have. This function is so important in certain markets that one could say that without certifiers "efficient trade would often be distorted, curtailed or blocked" (Barnett, 2012, 476). Against this background, it is not surprising that certification services are used in many sectors. Examples are the CE mark, audit opinions, the SGS product safety mark, the Underwriters Laboratories mark, food safety certificates, credit ratings, class certificates for vessels, energy performance certificates and certificates for medical devices (De Bruyne, 2019, 5).

Third-party certification is thus used as a regulatory tool to increase trust and ensure market efficiency. Yet, several scandals occurred illustrating that this certification mechanism does not always work efficiently. Credit rating agencies (CRAs), for instance, issued positive ratings to financial (structured) products that later defaulted, eventually leading to the 2008 financial crisis (Darbellay, 2013; Partnoy, 2017; De Bruyne, 2018). Likewise, classification societies certified the *Erika* and *Prestige* vessels that latter sank, thereby causing massive oil pollutions (Lagoni, 2007; De Bruyne, 2019). The PIP scandal also shows that notified bodies certified leaking breast implants that caused physical harm to women who purchased them (Verbruggen, 2018). In each of these cases, third-party certifiers issued certificates that (later) turned out to be unreliable and inaccurate. In some cases, third-party certifiers have even been held liable for the issuance of certificates that did not correspond to the 'actual' or 'real' value of the certified item (Hamdani, 2003; Partnoy, 2004; Coffee, 2004; De Bruyne, 2019). As such, academic scholarship has been focusing on the question which mechanisms could increase the accuracy and reliability of certificates. Much attention has, for instance, been given to reducing potential conflicts of interest between the certifier and the requesting entity, the creation of public or governmental-based certifiers or increasing competition in the certification sector (De Bruyne, 2019, 207–263 with references to these proposals).[3] More importantly, the 'deterring' effect or 'preventive' function of tort liability could also induce certifiers to issue more accurate and reliable certificates.

3.2 *Tort Liability as a Way to Increase the Accuracy and Reliability of Certificates*

There are several views on the role of tort law (Schwartz, 1997, 1801). In addition to corrective (e.g. Coleman, 1982; Weinrib, 1983 or distributive justice (Keating, 2000, 200), law and economics scholars understand tort law as an instrument aimed at the goal of deterrence (Calabresi, 1970; Posner, 1972). The purpose of damage payments in tort law is to provide incentives for potential injurers to take efficient cost-justified precautions to avoid causing the accident. An individual or entity makes the decision about whether or how to engage in a given activity by weighing the costs and benefits of the particular activity. The risk of liability and actual imposition of damages awards may lead parties to take into account externalities when they decide whether and how to act (Goldberg, 2003, 545). The fact that someone can be held liable *ex post* can thus provide the necessary incentives *ex ante* to act in such a way as to prevent liability (Faure and Hartlief, 2002, 19; Giesen, 2006, 14–15). The risk of having to bear financial burdens due to liability could serve as an incentive for potential tortfeasors to avoid injury-causing activities or at least to provide

3 Note, however, that these proposals have several legal challenges and shortcomings as well that need to be taken into account.

them with greater regard for safety (Brown, 1985, 976–977). Based on this reasoning, third-party certifiers will take into account – 'internalise' – the risk of civil liability when issuing their certificates of high-risk AI systems. This in turn will induce them to act more carefully, which could increase the accuracy and reliability of certificates (De Bruyne, 2019, 263–307). As profit maximisers, they can weigh the costs and benefits of their actions carefully and take a decision to prevent liability (Giesen, 2006, 16; Giesen, 2005, 148–149).

Zabinski and Black even find evidence "that reduced risk of med(ical) mal(practice) litigation, due to state adoption of damage caps, leads to higher rates of preventable adverse patient safety events in hospitals [...] Our study is the first, either for medical malpractice or indeed, in any area of personal injury liability, to find strong evidence consistent with classic tort law deterrence theory – liability for harm induces greater care" (2015, 19). Schwartz concludes that "tort law, while not as effective as economic models suggest, may still be somewhat successful in achieving its stated deterrence goals". More specifically, "[t]he information suggests that the strong form of the deterrence argument is in error. Yet it provides support for that argument in its moderate form: sector-by-sector, tort law provides something significant by way of deterrence" (1994, 443). Popper further argues that deterrence is a real and present virtue of the tort system. The actual or potential imposition of civil tort liability changes the behaviour of others. He concludes that "[a] tort case can communicate a normative message, an avoidance message, or a message affirming current practices. [footnote omitted] To deny that judicial decisions provide a valuable deterrent effect is to deny the historic role of the judiciary, not just as a matter of civil justice but as a primary and fundamental source of behavioral norms" (2012, 185). Based on a restricted survey, he even finds empirical evidence of the deterring function of tort law (2012, 196–197; De Bruyne, 2019a).

3.3 *Evaluating the Proposal for a Regulation on AI*

The deterring function or effect of tort law can thus be used as a mechanism to induce third-party certifiers to issue more accurate and reliable certificates for high-risk AI systems. Several factors, however, may undermine this deterring function (for a similar analysis for classification societies, see De Bruyne, 2019a). The question, for instance, arises as to whether certificates can be considered as opinions protected by the freedom of speech defence. This is important as the "Free Speech Clause [...] can serve as a defense in state tort suits" (Snyder v. Phelps, 562 US 443, 131 S.Ct. 1207, 1215 (2011)). Certifiers such as credit rating agencies have indeed already argued that their certificates should be protected under the constitutional or fundamental freedom of speech defence. Credit ratings are the "world's shortest editorials" written on an item or company's creditworthiness (Husisian, 1990, 446). CRAs, therefore, consider themselves as members of the financial press. As a consequence, ratings should be fully protected as journalistic speech (e.g. Abu Dhabi Commercial Bank v. Morgan Stanley & Co. Inc., 651 F. Supp. 2d 155, 175–176 (S.D.N.Y. 2009); In Re Enron Corp. Securities Derivative, 511 F.Supp.2d 742, 809–815 (S.D. Tex. 2005); Ohio Police & Fire v. Standard & Poor's Financial, 813 F.Supp.2d 871, 877 (2011)). Some courts – though surely not the vast majority – have in the past indeed considered this line of reasoning and qualified misleading ratings as predictive opinions not containing provably false factual connotations (e.g. Compuware Corp. v. Moody's Investors Services Inc., No. 05–1851, 23 August 2007, 7 (6th Cir. 2007); In Re Credit Suisse First Boston Corp., 431 F.3d 36, 47 at paragraph 32 (1st Cir. 2005); In Re Pan Am Corp., 161 B.R. 577, 586 (S.D.N.Y. 1993)). The rest of this section will briefly focus on two other important factors that may potentially reduce the deterring effect of tort law, namely potential conflicts of interests and the risk that certificates can become regulatory licenses. It will thereby also be examined whether and how these issues are tackled by the Proposal for a Regulation on AI.

3.3.1 Potential Conflicts of Interest

Conflicts of interest are often seen as a cause for unreliable and inaccurate certificates. A conflict of interest exists when a person in a certain situation has a duty to decide how to act solely based on the interests of another person while the choice he/she makes also has repercussions for his/her own interests (conflict of interest and duty) or for the interests of another, third person, that he/she is also legally bound to protect (conflict of duties). In some cases, a person is not only required to take into account the interests of certain parties when making its decision but also not allowed to consider the consequences of its choice towards the interests of other parties. This can occur when someone has an obligation to be impartial or make an independent judgment (Kruithof, 2008, 590–591, 595–596). A third-party certifier's duty to be independent and objective can conflict with its own interest on several moments during the entire certification process (conflict of interest and duty) (Davis, 1993, 21–41). A distinction can be made between two situations (De Bruyne, 2019, 215–231).

Conflicts of interest might, on the one hand, follow from the involvement of a third-party certifier in a requesting entity's activities. The certifier may assist the requesting entity in the design of the item that has to be certified. As a consequence, certifiers could make sure that the item is structured or designed in such a way that it will get a favourable certificate. Such conduct conflicts with the third-party certifier's required independence during the certification process (also see Art. R17 Decision 768/2008 on a common framework for the marketing of products). This conflict of interest has especially been at stake in cases dealing with the liability of CRAs after the 2008 financial crisis (e.g. Abu Dhabi Commercial Bank v. Morgan Stanley & Co., 651 F. Supp. 2d 155, 179 (S.D.N.Y. 2009)). Closely related to this conflict of interest is the situation in which a certifier provides auxiliary and consulting services to the requesting entity. A certifier may not only be involved in the certification process but can also offer other services to the entity. This could lead to the situation in which the certifier might be inclined to provide a favourable certificate to safeguard that it can offer additional services. Those additional services may increase the certifier's revenues but also impair its objectivity and independence (Crockett, 2003, 49; Sinclair, 2012, 122–124; De Bruyne, 2019, 216–219).

Conflicts of interest may, on the other hand, also relate to a third-party certifier's remuneration structure. Certifiers are paid by the requesting entity to issue the certificate. The question arises whether certifiers can in such circumstances really issue certificates in an independent way (Payne, 2015, 276–277). The undesired consequences resulting from this conflict of interest have already been identified for several certifiers. Deceived investors, for instance, claim that the so-called 'issuer-pays business model' caused CRAs to issue flawed ratings leading to the 2008 financial crisis (e.g. Anschutz Corporation v. Merill Lynch & Co. Inc., 785 F. Supp. 2d 799, 809 (N.D. Cal. 2011); Abu Dhabi Commercial Bank v. Morgan Stanley & Co. Inc., 651 F. Supp. 2d 155 (S.D.N.Y. 2009)). CRAs had to issue an independent rating on the issuer's creditworthiness or the latter's financial products while being paid by the very same issuer. Issuers may also threaten to contract with another CRA if the agency they contracted with uses standards that are too strict. CRAs may also issue 'unsolicited' ratings to safeguard that the issuer pays the necessary fees (Bai, 2010, 263–265). Such a conflict of interest can also occur during the conformity assessment procedure of medical devices, which involves notified bodies depending on the risks and class of the device. Manufacturers may contract with any of the notified bodies appointed by Member States for approval of their medical devices. They can even resubmit rejected applications to other bodies. At the same time, the scheme induces notified bodies to compete for business. There are many notified bodies providing their services within the EU.[4] Manufacturers can 'forum shop' for a notified body that is not too strict in the interpretation of supranational requirements or that offers the cheapest services. As a result, notified bodies might lower their standards to attract

4 See for an overview: https://ec.europa.eu/growth/tools-databases/nando/.

more manufacturers of medical devices (Fry, 2014, 174–176; Viana, 1996, 216; Chestnut, 2013, 341, De Bruyne, 219–220).

Art. 33 of the Proposal for a Regulation on AI also contains several provisions with regard to reducing potential conflicts of interest to ensure the issuance of accurate and reliable certificates of certain high-risk AI systems. The organisational structure, allocation of responsibilities, reporting lines and operation of notified bodies must ensure that there is confidence in the performance by and in the results of the conformity assessment activities that the body conducts. Notified bodies have to be independent of the provider of a high-risk AI system in relation to which it performs conformity assessment activities. They also need to be independent of any other operator having an economic interest in the high-risk AI system that is assessed, as well as of any competitors of the provider. Notified bodies have to be organised and operated so as to safeguard the independence, objectivity and impartiality of their activities. They need to document and implement a structure and procedures to safeguard impartiality and to promote and apply the principles of impartiality throughout their organisation, personnel and assessment activities. Reference can also be made to Decision 768/2008 on a common framework for the marketing of products, which outlines several requirements for notified bodies' independency in Art. R17. The requirements in the Proposal for a Regulation on AI as well as in Decision 768/2008, however, are rather general safeguards. They have often been further concretised in specific legislation covering third-party certifiers because their "independence is a key requirement for credible certification schemes" (Möslein and Zicari, 2021, 362). Based on a comparative analysis of the legal situation of other third-party certifiers, the following provisions could have been explicitly included in the Proposal to mitigate any potential conflicts of interest and hence increase the quality and reliability of certificates:[5]

– Specify a *maximum duration* of the contractual relationship between the provider of high-risk AI systems and a notified body.
  The idea is that multiple and different views, perspectives and methodologies applied by certifiers could produce more diverse certificates and ultimately improve the conformity assessment. This has been adopted for statutory auditors – i.e. certifiers that review the accuracy of a company's financial statements and issue an audit to express an opinion as to whether the financial statements fairly

5 Note, however, that these proposals have several legal challenges and shortcomings as well that need to be taken into account.

reflect, in all material aspects, the economic position of the company. Regulation 537/2014 on the requirements regarding statutory audits of public interest entities, for instance, includes a rotation mechanism for auditors. The Regulation is based on the premise that a maximum duration of the auditor's engagement strengthens his/her independence and hence the quality of audit opinions. A rotation mechanism has also been adopted in the context of CRAs. Regulation 462/2013 on CRAs introduces a mandatory rotation rule. With some exceptions (e.g. for small CRAs), the issuers of structured finance products with underlying re-securitised assets have to switch to a different CRA every four years. An outgoing CRA is not allowed to rate re-securitised products of the same issuer for a period equal to the duration of the expired contract, though not exceeding four years (Art. 6b Regulation 1060/2009 on CRAs as inserted by Art. 1(8) Regulation 462/2013).

– Specifically include that the notified body *cannot in any way be involved in the design and development* of high-risk AI systems.
  Pursuant to Art. R17 Decision 768/2008, notified bodies need to remain independent of the organisation or the product they assess. More specifically, they are not allowed to engage in any activity that may conflict with their independence of judgment or integrity such as consultancy services. A conformity assessment body, its top-level management and the personnel responsible for carrying out the conformity assessment tasks cannot be the designer, manufacturer, supplier, installer, purchaser, owner, user or maintainer of the products which they asses. This requirement is also included in legal frameworks applicable to other third-party certifiers such as notified bodies in the Medical Device Regulation. Notified bodies are not allowed to provide consultancy services to the manufacturer, a supplier or a commercial competitor on the design, construction, marketing or maintenance of medical devices. They cannot be involved in the design, manufacture or construction, marketing, installation and use or maintenance of devices they assess (Annex VII, Art.1.2 Regulation 2017/745 on medical devices). CRAs are also not allowed to provide consultancy or advisory services to the rated entity or any related third party regarding the corporate or legal structure, assets, liabilities or activities of that entity or related third party. Although a CRA may provide certain ancillary services other than issuing ratings (e.g. market forecasts, estimates of economic trends or pricing analysis), it must ensure that this does not create conflicts of interest with its rating activities (Annex I, B4, Regulation 1060/2009). Rating analysts

or persons who approve ratings are not allowed to make proposals or recommendations, either formally or informally, regarding the design of structured finance instruments on which the CRA is expected to issue a credit rating (Annex I, B5, Regulation 1060/2009; Darbellay, 2013, 32–33).

- Involve *two certifiers* or a *peer third-party certifier* (for certain) high-risk AI systems.

  One could also develop a scheme under which two certifiers need to issue a certificate for a particular item before it can be marketed. The involvement of a second certifier means that two additional 'eyes' are looking at the same item that needs to be certified. If the certificates for a particular item do not correspond, there might be an indication that one of the certifiers did not comply with its obligations during the certification process (Organisation for Economic Co-operation and Development, 2016, 79). This creates a higher risk of liability for a certifier and hence could increase the probability that it will issue certificates that are more reliable and accurate (Schweckendiek, van Tol and Pereboom, 2015, 46–77; Katzenbach, Leppla and Choudhury, 2016, 115). The idea of involving a second certifier is not entirely imaginary. In the context of CRAs, for instance, an issuer who intends to solicit a rating for a structured finance instrument has to appoint at least two CRAs to provide ratings independently from each other (Art. 8c Regulation 1060/2009 as inserted by Art. 1(11) Regulation 462/2013 on credit rating agencies). Alternatively, and more feasible, is to include a peer certifier. Whereas the 'primary' certifier remains responsible for issuing the certificate during the certification process, the peer certifier has to perform a review and submit a performance report to the public authority. This public authority could use this report for its existing supervisory and monitoring duties (see extensively discussing the merits as well as potential challenges De Bruyne, 2019, 358–373).

- Make sure that the *level of the remuneration* of those actors involved in the certification process does not depend on the results of the conformity assessment of high-risk AI systems

  Such provisions are included in legal frameworks related to other third-party certifiers as well. In the context of classification societies acting on behalf of flag States (i.e. Recognised Organisations-ROs), for instance, it is mentioned that they may not be substantially dependent on a single commercial enterprise for their revenue. The RO is not allowed to carry out class or statutory work if it is identical to or has business, personal or family links to the shipowner or operator. This incompatibility also applies to surveyors employed by an RO (Annex I, A, 6 Regulation 391/2009 on common rules and standards for ship inspection and survey organisations). Likewise, the level of the remuneration of the top-level management and assessment personnel of notified bodies may not depend on the results of the assessments of medical devices (Annex VII, Art. 1.2 Regulation 2017/745 on medical devices).

3.3.2 Certificates of High-Risk AI Systems as 'Regulatory Licenses'

Another factor that could potentially undermine the deterring function of tort law is caused by the regulatory use of certificates. Legislation often requires a certificate before the certified item can be marketed (cf. the 'gatekeeping function' of certifiers). Take the example of medical devices. A manufacturer can only place medical devices on the EU market that comply with the essential requirements. To that end, the manufacturer has to perform a conformity assessment procedure. The conformity assessment is conducted according to technical procedures included in legislation dealing with medical devices. The applicable legislation can require the involvement of notified bodies in the conformity assessment procedure. A notified body, for instance, needs to be involved in the conformity assessment procedure of certain high-risk medical devices. When such devices comply with the safety and technical criteria, the notified body will issue a certificate (Art. 52–60 and Annexes IX–XI of Regulation 2017/745 on medical devices). The manufacturer will only be able to market medical devices once the notified body has issued the required certificate.

The problem of extensively using and referring to certificates in legislation has especially been an issue in the context of CRAs. Regulators have to a certain extent "outsourced their safety judgments to third-party CRAs" (Darbellay, 2013, 40). As a consequence, CRAs shifted from selling information to selling "regulatory licenses" (Partnoy, 1999, 683), which are the "keys that unlock the financial markets" (Partnoy, 2009, 4). The use of extensive references to ratings in legislation has been criticised. It has, for instance, been argued that CRAs remain in business because financial legislation often requires a rating as a prerequisite for market access, for purchasing bonds by institutional investors or for other market activities, even if the rating turns out to be incorrect later (Partnoy, 2009, 4–6; Darbellay, 2013, 51–59). The fact that CRAs offer services that became necessary for regulatory compliance is one of the reasons that created and sustained the so-called 'paradox of credit ratings' (Partnoy, 2002, 65). This paradox implies that although the informational value of ratings decreases (e.g. because investors increasingly allege that CRAs issued flawed ratings), CRAs remain profitable and their ratings of major importance to regulate financial markets (Partnoy, 1999, 621–622).

Both EU and US regulators have since 2008 tried to eliminate references to or using ratings in legislation or other documents. The EU pursues the objective of reducing overreliance on ratings by adopting a multilayer approach, which *inter alia* implies that financial institutions are required to make their own credit risk assessment and not rely solely on ratings when assessing the creditworthiness of an entity or financial instrument. The EU recommended that legislation and supranational institutions should refrain from referring to ratings in their guidelines, recommendations and draft technical standards if it would cause authorities or other financial participants to rely 'solely' or 'mechanistically' on ratings (Directorate General Internal Market and Services, 2014, 4–5). In the US, the Dodd-Frank Act deals with the removal of references to ratings in a similar way. Section 939A directs each federal agency to review: (1) any regulation it has issued and which requires the use of an assessment of the creditworthiness of a security or money market instrument and (2) any references to or requirement of reliance on ratings in such regulations. Each agency has to modify any such regulations to remove the reference to or requirement of reliance on ratings (Section 939A (2010) Dodd–Frank Wall Street Reform and Consumer Protection Act, Pub. L. 111–203, 124 Stat. 1376–2223).

The Proposal for a Regulation on AI does not seem to address this issue. It should, nevertheless, be considered in the context of certifying high-risk AI systems. It has to be ensured that the future regulatory framework does not solely or mechanistically rely on or include certificates given to high-risk AI systems to prevent third-party certifiers from issuing 'regulatory licenses'. As such, a (difficult) balance needs to be established between certification as a regulatory tool for market access of high-risk AI systems and the risk of overreliance on those certificates in legal frameworks, potentially resulting in the previously mentioned paradox. That being said, the Proposal does contain other safeguards regarding the value of certificates. Art. 44 provides that certificates are valid for the period they indicate, which may not exceed five years. On application by the provider, the validity of a certificate may be extended for further periods, each not exceeding five years, based on a re-assessment in accordance with the applicable conformity assessment procedures. When a notified body finds that an AI system no longer meets the applicable requirements, it has to suspend or withdraw the issued certificate or impose any restrictions on it unless compliance with those requirements is ensured by appropriate corrective action taken by the provider of the system within an appropriate deadline set by the notified body. The notified body needs to give reasons for its decision.

4 Concluding Remarks

This contribution shed light on some considerations with regard to the third-party certification of AI. It started by giving an overview of some regulatory and policy initiatives on AI in the EU. Particular attention was thereby given to several important provisions in the Proposal for a Regulation on AI. It was shown that third-party certification will play an important role for certain high-risk AI systems and especially that the risk of incurring liability may have a so-called 'deterring' effect, hence increasing the accuracy and reliability of certificates. This in turn may potentially enhance the public's trust in artificial intelligence. The chapter illustrated the importance of additional research on third-party certification and AI. More in general, the use and role of (tort) liability as a trust-enhancing mechanism needs more consideration in light of the AI (r)evolution.

Acknowledgement

This chapter is based on previous research on third-party certifiers (De Bruyne, 2019) and will be relied upon to conduct more extensive research on third-party certification and AI. I would like to thank Orian Dheu for his valuable feedback on earlier versions of this work. It should be noted that this chapter is mainly based on existing literature and policy initiatives available at the time of the conference in 2021.

References

Scholarship

Abbott, R. (2020). *The Reasonable Robot: Artificial Intelligence and the Law*. Cambridge: Cambridge University Press. 300 p.

Akerlof, G.A. (1970). The Market for "Lemons": Quality Uncertainty and the Market Mechanism. *The Quarterly Journal of Economics*, 84, 488–500.

Bai, L. (2010). On Regulating Conflict of Interests in the Credit Rating Industry. *New York University Journal of Legislation and Public Policy*, 13, 253–313.

Barfield, W., and Pagallo, U. (2018). *Research Handbook on the Law of Artificial Intelligence*. Cheltenham: Edward Elgar Publishing. 736 p.

Barnett, J. (2012). Intermediaries Revisited: Is Efficient Certification Consistent with Profit Maximization?. *Journal of Corporation Law*, 37, 476–522.

Brijs, T. (2021). Verkeersveiligheid. In: De Bruyne J. (Ed.), *Autonome motorvoertuigen. Een multidisciplinair onderzoek naar de maatschappelijke impact* (127–147). Brugge: Vanden Broele.

Brown, C. (1985). Deterrence in Tort and No-Fault: The New Zealand Experience. *California Law Review*, 73, 976–1002.

Buri, T., and von Bothmer, F. (2021). The New EU Legislation on Artificial Intelligence: A Primer. SSRN paper. Retrieved from https://papers.ssrn.com/sol3/papers.cfm?abstract_id=3831424.

Calabresi, G. (1970). *The Costs of Accidents: A Legal and Economic Analysis*. New Haven: Yale University Press. 340 p.

Calo, R. (2016). Robots in American Law. University of Washington School of Law Research Paper no. 2016–04. 45 p.

Chestnut, B. (2013). "Cherry" Trees or "Lemon" Trees: Conflicts of Interest in Forest Certification. *Georgetown International Environmental Law Review*, 25, 341–366.

Coeckelbergh, M. (2020). *AI Ethics*. Cambridge: MIT Press. 248 p.

Coffee, J.C. (2004). Gatekeeper Failure and Reform: The Challenge of Fashioning Relevant Reforms. *Boston University Law Review*, 84, 301–364.

Coleman, J. (1982). Moral Theories of Torts: Their Scope and Limits: Part I. *Law and Philosophy*, 1, 371–390.

Crockett, A. (2003). *Conflicts of Interest in the Financial Services Industry: What Should We Do about Them?*, Geneva: Centre for Economic Policy Research. 100 p.

Custers, B. and Fosch Villaronga, E. (2022). *Law and Artificial Intelligence. Regulating AI and Applying AI in Legal Practice*. Berlin: Springer. 569 p.

Darbellay, A. (2013). *Regulating Credit Rating Agencies*. Cheltenham: Edward Elgar. 296 p.

Davis, M. (1993). Conflict of Interest Revisited. *Business and Professional Ethics Journal*, 12, 21–41.

De Bruyne, J., and Tanghe, J. (2017). Liability for damage caused by autonomous vehicles: a Belgian perspective. Journal of European Tort Law, 8, 324–371.

De Bruyne, J. (2018). A European perspective on the liability of credit rating agencies. *Journal of International Business and Law*, 17, 233–253.

De Bruyne, J. (2019). *Third-Party Certifiers*. London: Kluwer Law International. 464 p.

De Bruyne, J. (2019a). Tort law and the regulation of classification societies: between public and private roles in the maritime industry. *European Review of Private Law*, 27, 429–450.

De Bruyne, J., and Vanleenhove, C. (2022). *Artificial intelligence and the Law*. Antwerp: Intersentia. 668 p.

De Bruyne, J., Bouteca, N., (Eds.). (2021). *Artificiële intelligentie en maatschappij*. Turnhout: Gompel & Svacina. 337 p.

De Bruyne, J., Van Gool, E., and Gils, T. (2021). Tort Law and Damage Caused by AI System. In: De Bruyne, J., and Vanleenhove, C., (Eds.), *Artificial intelligence and the Law* (359–405). Antwerp: Intersentia.

Dignum, V. (2019). *Responsible AI*. Cham: Springer. 127 p.

Dubber, M.D., Pasquale, F., and Das, S. (2020). *Oxford Handbook of Ethics of AI*. Oxford: Oxford University Press. 1000 p.

Ebers, M et al. (2021). The European Commission's Proposal for an Artificial Intelligence Act – A Critical Assessment by Members of the Robotics and AI Law Society (RAILS). Multidisciplinary Scientific Journal, 4, 4, 589–603.

Ebers, M. (2022). Standardizing AI – The Case of the European Commission's Proposal for an Artificial Intelligence Act. In: DiMatteo L.A. et al. (Eds.), *The Cambridge Handbook of Artificial Intelligence* (321–344). Cambridge: CUP.

Faure, M.G., and Hartlief, T. (2002). *Nieuwe risico's en vragen van aansprakelijkheid en verzekering*. Deventer: Kluwer. 338 p.

Fry, B.M. (2014). A Reasoned Proposition to a Perilous Problem: Creating a Government Agency to Remedy the Emphatic Failure of Notified Bodies in the Medical Device Industry. *Willamette Journal of International Law & Dispute Resolution*, 22, 161–198.

Garner, B. (2009). *Black's Law Dictionary*. St. Paul: West. 1920 p.

Greenberg, D. (2010). *Jowitt's Dictionary of English Law*. London: Sweet & Maxwell. 2496 p.

Goldberg, J.C.P. (2003). Twentieth-Century Tort Theory. *Georgetown Law Journal*, 91, 513–584.

Giesen, I. (2005). *Toezicht en aansprakelijkheid: een rechtsvergelijkend onderzoek naar de rechtvaardiging voor de aansprakelijkheid uit onrechtmatige daad van toezichthouders ten opzichte van derden*. Deventer: Kluwer. 251 p.

Giesen, I. (2006). Regulating Regulators Through Liability – The Case for Applying Normal Tort Rules to Supervisors. *Utrecht Law Review*, 2, 8–31.

Hamdani, A. (2003). Gatekeeper Liability. *Southern California Law Review*, 77, 53–122.

Hawes, C., Hatzel, J., and Holder, C. (5 May 2021). The Commission's proposed Artificial Intelligence Regulation. Retrieved from: https://www.bristows.com/news/the-commissions-proposed-artificial-intelligence-regulation/.

Hervey, M. (Author), Lavy, M. (2020). *The Law of Artificial Intelligence Hardcover*. London: Sweet & Maxwell. 648 p.

Husisian, G. (1990). What Standard of Care Should Govern the World's Shortest Editorials? An Analysis of Bond Rating Agency Liability. *Cornell Law Review*, 75, 410–460.

Ivanov, S.H. (2017). Robonomics – Principles, Benefits, Challenges, Solutions. Yearbook of Varna University of Management, 10, 283–293.

Katzenbach, R., Leppla, S., and Choudhury, D. (2016). *Foundation Systems for High-Rise Structures*. London: CRC Press. 298 p.

Keating, G.C. (2000). Distributive and Corrective Justice in the Tort Law of Accidents. *Southern California Law Review*, 74, 193–224.

Kruithof, M. (2008). Wanneer vormen tegenstrijdige belangen een belangenconflict?. In: Van der Elst, C.,De Wulf, H., Steennot, R., and Tison, M. (Eds.), *Van alle markten: liber amicorum Eddy Wymeersch* (575–598). Antwerp: Intersentia.

Lagoni, N. (2007). *The Liability of Classification Societies*. Berlin: Springer. 380 p.

Lofgren, K.-G., Persson, T., and Weibull, J.W. (2002). Markets with Asymmetric Information: The Contributions of George Akerlof, Michael Spence and Joseph Stiglit. *Scandinavian Journal of Economics*, 104, 195–211.

Mannens, E. (2021). Wat je moet weten over AI. In: De Bruyne, J. and Bouteca, N. (Eds.), *Artificiële intelligentie en maatschappij* (17–49). Turnhout: Gompel & Svacina.

Möslein, F., and Zicari, R. (2021). Certifying artificial intelligence systems. In: Volg, R. (Ed.), *Research Handbook on Big Data Law* (357–374). Cheltenham: Edward Elgar Publishing, 2021.

Partnoy, F. (1999). The Siskel and Ebert of Financial Markets: Two Thumbs Down for the Credit Rating Agencies. *Washington University Law Quarterly*, 77, 619–714.

Partnoy, F. (2002). The Paradox of Credit Rating. In: Levich, R.M., Majnoni, G. and Reinhart, C. (Eds.), *Ratings, Rating Agencies and the Global Financial System* (65–97). New York: Springer.

Partnoy F., (2004). Strict Liability for Gatekeepers: A Reply to Professor Coffee. *Boston University Law Review*, 84, 365–375.

Partnoy, F. (2009). Rethinking Regulation of Credit Rating Agencies: An Institutional Investor Perspective. *Legal Studies Research Paper Series Research Paper No. 09-014*.

Partnoy, F. (2017). What's (Still) Wrong with Credit Ratings. *Washington Law Review*, 92, 1407–1472.

Payne, J. (2015). The Role of Gatekeepers. In: Ferran, E., Moloney, N. and Payne, J. (Eds.), *The Oxford Handbook of Financial Regulation*. Oxford: Oxford University Press. 795 p.

Popper, A.F. (2012). In Defense of Deterrence. *Albany Law Review*, 75, 181–204.

Posner, R.A. (1972). A Theory of Negligence. *Journal of Legal Studies*, 1, 29–96.

Schwartz, G.T. (1997). Mixed Theories of Tort Law: Affirming Both Deterrence and Corrective Justice. *Texas Law Review*, 75, 1801–1834.

Schwartz, G.T. (1994). Reality in the Economic Analysis of Tort Law: Does Tort Law Really Deter. *UCLA Law Review*, 42, 377–443.

Schweckendiek, T., van Tol, A.F. and Pereboom, D. (2015). *Geotechnical Safety and Risk*. Amsterdam: IOS Press. 1028 p.

Sinclair, T.J. (2012). The Credit Rating Enigma. *Global Dispatches*, 18 December 2012.

Smuha, N. et al. (2021). How the EU Can Achieve Legally Trustworthy AI: A Response to the European Commission's Proposal for an Artificial Intelligence Act. SSRN Paper. 64 p.

Tzafestas, S.G. (2015). *Roboethics: A Navigating Overview*. Athens: Springer. 204 p.

Veale, M. and Borgesius Zuiderveen, F. (2021). Demystifying the Draft EU Artificial Intelligence Act. *Computer Law Review International,* 4, 97–112.

Verbruggen, P. (2013). Aansprakelijkheid van certificatie-instellingen als private toezichthouder. *Nederlands Tijdschrift voor Burgerlijk Recht,* 30, 329–337.

Verbruggen, P., and Van Leeuwen, B. (2018). The Liability of Notified Bodies under the EU's New Approach: The Implications of the PIP Breast Implants Case. *European Law Review*, 43, 394–409.

Viana, V.M. (1996). *Certification of Forest Products: Issues and Perspectives*. Washington: Island Press. 261 p.

Weinrib, E.J. (1983). Toward a Moral Theory of Negligence Law. *Law and Philosophy*, 2, 37–62.

Zabinski, Z. and Black, B.S. (2015). The Deterrent Effect of Tort Law: Evidence from Medical Malpractice Reform. *Northwestern Law & Econ Research Paper No. 13–09*.

Case Law

Abu Dhabi Commercial Bank v. Morgan Stanley & Co. Inc., 651 F. Supp. 2d 155 (S.D.N.Y. 2009).

Anschutz Corporation v. Merill Lynch & Co. Inc., 785 F. Supp. 2d 799 (N.D. Cal. 2011).

Compuware Corp. v. Moody's Investors Services Inc., No. 05-1851, 23 August 2007 (6th Cir. 2007).

In Re Enron Corp. Securities Derivative, 511 F.Supp.2d 742 (S.D. Tex. 2005).

In Re Credit Suisse First Boston Corp., 431 F.3d 36 (1st Cir. 2005).

In Re Pan Am Corp., 161 B.R. 577 (S.D.N.Y. 1993).

Ohio Police & Fire v. Standard & Poor's Financial, 813 F.Supp.2d 871 (2011).

Snyder v. Phelps, 562 U.S. 443, 131 S.Ct. 1207 (2011).

Legislation

Decision 768/2008 on a common framework for the marketing of products, OJ L 218.

Dodd–Frank Wall Street Reform and Consumer Protection Act, Pub. L. 111–203, 124 Stat. 1376–2223.

Regulation (EC) No 1060/2009 of the European Parliament and of the Council of 16 September 2009 on credit rating agencies, OJ L 302.

Regulation (EU) No 462/2013 of the European Parliament and of the Council of 21 May 2013 amending Regulation (EC) No 1060/2009 on credit rating agencies, OJ L 146.

Regulation (EU) No 537/2014 of the European Parliament and of the Council of 16 April 2014 on specific requirements regarding statutory audit of public-interest entities and repealing Commission Decision 2005/909/EC, OJ L 158.

Regulation (EU) 2017/745 of the European Parliament and of the Council of 5 April 2017 on medical devices, amending Directive 2001/83/EC, Regulation (EC) No 178/2002 and Regulation (EC) No 1223/2009 and repealing Council Directives 90/385/EEC and 93/42/EEC, OJ L 117.

Regulation (EC) No 391/2009 of the European Parliament and of the Council of 23 April 2009 on common rules and standards for ship inspection and survey organisations, OJ L 131.

Other

AI HLEG, "Ethics Guidelines for Trustworthy AI", 8 April 2019.

AI HLEG, High Level Expert Group on AI, "Assessment List for Trustworthy Artificial Intelligence (ALTAI) for self-assessment", 17 July 2020.

Directorate General Internal Market and Services, "Staff Working Paper, EU Response to the Financial Stability Board: EU Action Plan to Reduce Reliance on Credit Rating Agency (CRA) Ratings", 12 May 12, 2014.

European Commission, Press Release, IP/19/1893, "Artificial intelligence: Commission takes forward its work on ethics guidelines", 8 April 2019.

European Commission, "Communication from the Commission to the European Parliament, the European Council, the Council, the European Economic and Social Committee and the Committee of the Regions on Artificial Intelligence for Europe", 25 April 2018.

European Commission, "White Paper On Artificial Intelligence – A European approach to excellence and trust", COM(2020) 65 final, 19 February 2020.

Organisation for Economic Co-operation and Development, "OECD Public Governance Reviews Integrity Framework for Public Investment", OECD Publishing, 2016.

Organisation for Economic Co-operation and Development, "Artificial Intelligence in Society", 11 June 2019.

CHAPTER 5

AI in 5G: High-Speed Dilution of the Right to Privacy and Reducing the Role of Consent

Mykyta Petik

The telecommunications industry has gradually migrated from legacy telephony networks to telephony networks based on an Internet Protocol (IP) network to deliver communication services. (Zhang, 2018) It is projected that by 2025, 5th generation of mobile networks (5G) will occupy 20% of the market share of mobile telecommunication technologies worldwide (O'Dea, 2020). The majority of Europeans use smartphones and share their sensitive personal data via mobile networks. This reinforces the notion of humans being 'always connected' due to the large-scale utilization of modern technologies and inherent challenges to privacy it entails (Madhusanka, et al., 2018). 5G takes advantage of numerous technological novelties to provide lower latency and faster data processing. Artificial intelligence (AI), Machine Learning (ML), and cloud-based solutions bring new actors into the EU mobile networks, change the way networks are secured, affect how privacy is protected, and personal data is processed.

Against this backdrop, new legal research is needed to fill the gap between the deployment of AI in 5G networks and the legal mechanisms behind the use of such technology aimed at protection of privacy and personal data. AI and ML may prove to become both enablers of privacy-preserving solutions and significant privacy disruptors. It comes as no surprise that the emergence of novel technologies like AI and 5G is associated with the challenges to privacy and balance of interests. Given the recent changes in data protection regulation and the evolving case law in this regard (Case C-131/12 ("Google Spain") and Case C-311/18 ("Schrems II")), the issues of privacy and data protection have become a prime factor in the modern digital economy.

This chapter will analyze the complex interplay between two rapidly evolving technologies – AI and 5G – and the legal concepts of privacy and consent to processing personal data. It is not the intention here to deliberate over a substantial volume of legal, ethical, economic, and societal challenges pertaining to the use of the abovementioned technologies but rather concentrate on how the AI and 5G brought together threaten to dilute the legal boundaries of

 | DOI:10.1163/9789004547261_007

privacy and reduce the importance of the consent to the processing of personal data.

1 Privacy and Data Protection: Clarifying the Setting

The legal discussion about privacy in relation to an individual's home, reputation, and written communication existed for centuries, with some legal cases concerning eavesdropping dating back to 14th century England (Holvast, 2009). Privacy as a right had a long way to be formed and universally recognized, though. More than 130 years ago, American lawyers S. Warren and L. Brandeis published an article titled 'The Right to Privacy', which greatly contributed to recognizing and developing the new legal right of the same name. "*Political, social, and economic changes entail the recognition of new rights*", as stated by the authors in the introduction to the article, was not a newly invented idea but served as a foundation for the argument for the recognition of the right to privacy (Warren & Brandeis, 1890).

Privacy was mainly associated with the integrity of the reputation and protection of home, personal life and communications and interpreted as "*the right to be let alone*" until the 1960s. It was then that the breakthrough in technological development – computers and their use for the storage of data – reignited the debate over privacy as a legal right. Early computerization and automated systems in the 1960s made the collection and transfer of personal data much easier, which prompted first discussions over the need for the legal regulation of the processing of personal data (Kosta, 2013). Technology caused us to re-think the approaches to understanding privacy. The concept of privacy as "the right to be let alone" gradually shifted towards the right to exercise control over the information recorded about a particular individual (Kosta, 2013).

Through numerous discussions and legislative developments in the 20th century, the right to privacy has become an accepted concept in Europe. The process was not universal and straightforward, though. It took various working groups and a board of experts from the Council of Europe, OECD, European Communities, and different national states to develop multiple legislation extending the application of the right to privacy to the new legal relationships caused by technological developments. The European Court of Human Rights (ECtHR) has adopted a dynamic and evolutive interpretation of the concepts of 'private life', which was stated to have a broad character "*incapable of exhaustive definition*", and 'personal interests', in order to produce decisions in line with technological advances (European Court of Human Rights, 2020). Moreover, in several cases (Copland v. the United Kingdom, Bărbulescu

v. Romania, Wieser and Bicos Beteiligungen GmbH v. Austria, Petri Sallinen and Others v. Finland, Iliya Stefanov v. Bulgaria) the ECtHR asserted that technologies come within the scope of Article 8 of the European Convention on Human Rights (ECHR) (European Court of Human Rights, 2020).

Privacy, being a fundamental right according to both the ECHR and the Charter of Fundamental Rights of the European Union (CFREU) is aimed to protect the body, family, home, and communication of the individual (European Court of Human Rights, 2020). The above-mentioned article by Samuel Warren and Louis Brandeis raised the issue of privacy in the context of newspaper publications and gossip, which may harm the individuals' reputation and psychological well-being. Recent technological advances such as the massive use of mobile networks for storage of data, AI, and surveillance provided grounds for large scale invasion of privacy and the need to impose stricter rules on handling personal data. Implementation of 5G networks will have a dramatic impact not only on the economy and society, but also on the legal safeguards and protection of privacy and personal data.

The adoption of the General Data Protection Regulation (GDPR) by the European Union and its entry into force in May 2018 recognized the protection of personal data as a fundamental right distinguishing it from the right to privacy, guaranteed by the ECHR and CFREU. The GDPR serves as the general act on the protection of natural persons with regard to the processing of their personal data and applies, among other things, to mobile networks. The GDPR was adopted as a technology-neutral, risk-based, and future-proof law based on principles and deals primarily with the legal basis, aims, instruments, and limits regarding the processing of personal data and individuals' rights to the protection of their personal data.

The concept of personal data is particularly important since it essentially defines the scope and aim of the whole regulation. According to Article 4(1) of GDPR personal data is "any information relating to an identified or identifiable natural person ('data subject'); an identifiable natural person is one who can be identified, directly or indirectly, in particular by reference to an identifier such as a name, an identification number, location data, an online identifier or to one or more factors specific to the physical, physiological, genetic, mental, economic, cultural or social identity of that natural person".

The European Commission stated that the implementation of new technologies does not put the foundational principles of GDPR into question but rather affects the application of these principles to specific technologies (European Commission, 2020). Unlike the previous generation of mobile technologies such as 4G, a wide range of principles and rights established by the GDPR, such as privacy by design, right to be informed, right to rectification, and right to object will have a substantial effect on the implementation of 5G networks and AI in the EU (Rizou, Alexandropoulou-Egyptiadou, & Psannis, 2020). The GDPR sets several grounds for the legitimate processing of personal data, one of which is the data subject's consent. Consent, in general, is a complex legal and ethical concept giving legitimacy to specific actions regarding the individual or their property and is a core principle of the system of protection of human rights. Article 7 GDPR states that consent must be given freely, specific, informed, unambiguous, and the data controller can demonstrate that the data subject has consented to the processing of their personal data. The request for consent must be clearly distinguishable from other information or requests and use clear and plain language. In this chapter, we interpret consent as an affirmative action concerning the processing of personal data following the GDPR meaning. In the context of EU law, data protection and privacy are treated as separate concepts, with discussion ongoing on their precise legal nature.[1] Despite considerable overlapping, the legal nature of data protection and privacy differs in their scope.

2 AI in 5G Networks: Disruptive Novelties and the Role of Consent

Being an evolution of previous mobile network technology, 5G offers substantially higher data transfer rates, significantly shorter network latency, and greater bandwidth, allowing more significant download and upload speeds and a broad spectrum of use cases. 5G shall become a major enabler for the Internet of Things ('IoT'), which paves the way for among other things, remote surgery, telemedicine, autonomous cars, smart homes, offices, warehouses, highways, and eventually the entire smart cities ecosystem (Blackman & Forge, 2016). Likewise, the IoT technologies will transform sectors that are not traditionally associated with Internet technologies, such as agricultural, automotive, and construction sectors. The innovation brought by 5G shall become possible by deploying millions IoT devices (sensors, trackers, smartphones,

1 See, for instance, Serge Gutwirth and Paul De Hert, 'Privacy, Data Protection and Law Enforcement. Opacity of the Individual and Transparency of Power' in E Claes, A Duff and S Gutwirth (eds), *Privacy and the Criminal Law* (Intersentia 2006); Gloria González Fuster and Raphaël Gellert, 'The Fundamental Right of Data Protection in the European Union: In Search of an Uncharted Right' (2012) 26 International Review of Law, Computers & Technology 73; Gloria González Fuster and Serge Gutwirth, 'Opening up Personal Data Protection: A Conceptual Controversy' (2013) 29 Computer Law & Security Review 531; Orla Lynskey, 'Deconstructing Data Protection: The "Added Value" Of A Right To Data Protection In The EU Legal Order' (2014) 63 International and Comparative Law Quarterly 569;.

home appliances), generating massive amounts of data. 5G enables fast transfer of this data to the computing facilities, where it is analyzed, and the response is carried back to the device, hence to the end-user.

5G networks rely partially on previous generations of mobile networks' technologies and infrastructure, namely 4G and 3G. Is it possible to say that considering all the technological novelties, 5G is not just succeeding 4G as the next generation of mobile networks but paves the way for a completely new type of mobile network.[2] Moving forward, Higher reliance on Internet-based services and a drastic increase in cellular data volume growth forces network operators and service providers to change existing approaches to establishing and maintaining core networks by paying more attention to security and efficiency. Modern telecommunication service providers face several challenges related to growth in mobile data exchange and in a number of devices such as increasing operational costs, the necessity to accommodate the coexistence of different generations of cellular networks such as 3G, 4G, and 5G, the need to upgrade equipment and the need to expand network capacity to accommodate an unprecedented number of Internet-connected devices. The pressure on existing mobile networks shall grow because of rising numbers of personal smart mobile devices and industrial Internet of Things (IoT) devices, with research claiming that Machine-to-Machine (M2M) communications shall generate larger amounts of data than all data generated by humans combined (Zhang, 2018). The expansion of physical infrastructure alone is not enough to keep up with the changes in mobile technologies and their usage. To cope with the challenge of processing immense data, there is a need to use advanced and efficient algorithms routing the data through the mobile network. Researchers indicate that to become fully operational and efficient, 5G networks must employ AI systems (Morocho Cayamcela & Lim, 2018). The security and privacy architecture of 5G networks will be much more complex and multilayered than previous generations of mobile networks. These requirements arise from the immense amount of data processed in 5G core network. AI and ML solutions have been used before for modeling attack patterns to enhance security and privacy against sophisticated cyberattacks and hacks. This will most likely be the case for 5G network use cases as well. AI and ML may help

2 *"Because 5G is not the same as the Gs that came before. Implementing this technology does not simply require an upgrade of the current network, it demands a new type of network altogether"*. per Ivezic, M. (2020). Unlocking the Future – Why Virtualization Is the Key to 5G. 5G Security – 5G, MIoT, CPSSEC, Security Blog by Marin Ivezic. https://5g.security/5g/virtualization-key-5g/.

identify the weaknesses in critical security points in authentication, identity, and assurance layers (Haider, Zeeshan, & Imran, 2020).

High speed of data processing in 5G networks enabled by AI and ML implementation may lead to practical problems in obtaining data subject's consent, providing information on processing, informing authorities, and affected data subjects on data breaches as well as concerns regarding interoperability and the execution of the right to erasure and rectification of personal data (Rizou, Alexandropoulou-Egyptiadou, & Psannis, 2020). 5G networks paired with AI algorithms will contribute to gathering and processing unprecedented amounts of personal data, which may require stricter standards of handling and processing, and will be transferred and processed continuously.

But with high speed comes a challenge to privacy. It would be tough to obtain an individual's valid consent for processing their personal data by AI and ML in 5G networks before such processing takes place because of the almost simultaneous transfer and processing of substantial amounts of personal data. A valid consent is a legal basis for processing of personal data and is mandatory in certain cases. According to Article 7 GDPR "[t]he data subject shall have the right to withdraw his or her consent at any time. The withdrawal of consent shall not affect the lawfulness of processing based on consent before its withdrawal. Prior to giving consent, the data subject shall be informed thereof. It shall be as easy to withdraw as to give consent". As was pointed out by Eleni Kosta, "The role of consent in this era is reduced, as the control of the individual over his personal information is overcome by the facilitation of everyday activities in electronic communications and especially the internet, to the extent that the privacy of the individual is not infringed" (Kosta, 2013). Extensive information about the devices and wearables of the data subject including sensitive health data such as pulse rate or fitness will be gathered and processed using AI almost instantly without the possibility to obtain valid consent the moment such data subject is within the range of 5G-based IoT systems. While data subjects may provide their consent to gather and process such data locally or for a specific purpose (e.g. tracking the pulse or other health data), the use of such data for training ML algorithms may be considered too extensive. According to 'A first assessment based on stakeholder inputs – Report on the impact of 5G on regulation and the role of regulation in enabling the 5G ecosystem' by the Body of European Regulators for Electronic Communications (BEREC) there is a risk of inability to obtain a valid consent by multiple data processing actors in 5G value chain since such actors will not have a direct relation to end-users (BEREC, 2019). Another significant challenge is the feasibility of verifying child's consent to data processing. According to Article 8 of GDPR "Where the child is below the age of 16 years, such processing shall be lawful only if and to the extent that consent is given or authorised by the holder of parental responsibility over the child" and "The controller shall make reasonable efforts to verify in such cases that consent is given or authorised by the holder of parental responsibility over the child, taking into consideration available technology". The data controller has an obligation to undertake 'reasonable efforts' to verify whether child's parents or caretakers indeed provided their consent to processing child's personal data.

AI and ML solutions will also enable real-time profiling and individual automated decision-making regulated by Article 22 GDPR. The Article states that the right of the data subject bot to be subject to a decision based exclusively on automated processing unless such profiling: (a) is necessary for entering into or performing a contract between a data subject and a controller; (b) is authorized by EU or Member State law; (c) is based on the explicit consent of the data subject. The last requirement is the toughest to comply with, considering the problem of obtaining data subject's consent before the processing of data occurs. Article 7(2) GDPR directly requires to inform the data subject of the content of consent *before* the data subject can give it.

5G relies on a larger number of base stations, IoT devices and sensors than previous generations of mobile networks. Precise geolocation of data subjects assisted by AI and ML algorithms under such conditions becomes much more accessible. In its Opinion 13/2011 on Geolocation services on smart mobile devices Article 29 Working Party (Article 29 WP, replaced by the European Data Protection Board after the adoption of GDPR) stated that "*smart mobile device is very intimately linked to a specific individual*" (Article 29 Working Party, 2011). The owner of the smartphone stores large amounts of personal data on the device and is unlikely to give it to another person for a considerable amount of time, which allows to use a smartphone as a unique identifier of the individual. Anyone who gets lawful or unlawful access to the geolocation data may learn the home address, workplace, habits, and routine of the data subject. Geolocation data may be qualified as a special category of personal data if it reveals political opinions (e.g. attendance of political rallies, gatherings, meetings), religious beliefs (attendance of religious establishments or sites), trade union membership (attendance of trade union meetings) of the data subject as stated in Article 9(1) of GDPR. While geolocation was possible in 3G and 4G networks, 5G networks bring it to another level by using AI and ML on a larger scale and enabling the high precision and speed of such geolocation not possible in previous mobile networks. Correspondingly, the requirements for informed consent may be stricter since the impact of the data breach or data leak is higher.

Obtaining an informed and freely given consent is a major challenge, since it might be hard to articulate all required information properly and accurately regarding the technology involved (AI), specific aims of data processing, and timeframe of storage of data, among other things. Technical issues deepen this challenge, as data controller must obtain consent before the processing actually takes place, which is not always realistic in the setting of AI in 5G networks. Possibly, solution providers will provide very limited information in the consent and ask for it *as hoc*, after processing personal data they only ask the consent for. These factors greatly contribute to reducing the role of consent in 5G networks.

3 The Impact of AI in 5G Networks on Privacy and Principles Relating to the Processing of Personal Data

The issue of possible adverse effects on privacy posed by AI implementation was addressed by the Parliamentary Assembly of the Council of Europe in its Recommendation on Technological convergence, artificial intelligence, and human rights. The Recommendation recognized the possibility of consequences for human rights in light of the intersection between various technology sectors, including AI solutions (Parliamentary Assembly, 2017).

AI and ML were described as *"perfect match"* for the use in 5G networks since AI algorithms can learn from complex patterns and assist with autonomous operations within the network, creating a scalable and data-driven solution (IEEE Future Networks World Forum 2021). AI can be used for developing 5G network architecture, automating functions such as optimization, security analysis, and fault prediction. Both AI and ML algorithms will analyze user's data transferred by 5G networks to develop more efficient services and applications. An important feature of AI in 5G is that it is used in all layers – from Radio Access Network (RAN) to core network and edge computing layer. AI is not limited to specific operations but may be rather omnipresent in 5G. In relation to privacy and protection of personal data AI and ML solutions can be deployed to optimize the use of resources for data processing in the cloud, detect anomalies and generate predictive analytics based on user's data. To train the AI algorithms and obtain useful and accurate results, large data must be fed to the algorithm. The more precise and up-to-date data is provided, the more tailored AI algorithm is obtained. Under normal circumstances developing a working AI would take years and the large number of resources. 5G network may serve as a great source of data for training AI algorithms since virtualization and edge computing technology significantly contribute to practical storage and management of ever-expanding human-generated data. In a context where AI technology is deployed as a part of the larger 5G ecosystem, there is a need to protect individuals' rights to privacy and protection of personal data while simultaneously refraining from unnecessarily hindering technological development.

One of the crucial challenges is determining the statuses of (joint) controllers and processors in 5G networks. Data controller is a natural or legal person or public authority or agency that determines the purposes and means of processing personal data, and data processor is any entity that processes this personal data on behalf of the controller. Joint controllers jointly determine the purposes and means of processing personal data, according to Article 26 of GDPR. However, the controller's status is autonomous, and the allocation of legal responsibilities must be where the real influence on data processing-related decisions is. AI and ML solutions can be provided by multiple companies and individuals from different jurisdictions, possibly involving non-EU states. This complexity causes the need for consideration of cross-border transfers of data, establishing effective data governance, resolving the problem of data ownership, and ensuring transparency.

AI and ML algorithms used in 5G networks will also have to comply with the principles relating to processing personal data described in Article 5 GDPR which serve as a foundation of protecting personal data in the EU. These principles can be applied in different contexts and are not limited to a particular technical, social, or economic domain (Kuner, Bygrave, Docksey, & Drechsler, 2020). The principles include:

- Lawfulness, fairness, and transparency;
- Purpose limitation;
- Data minimization;
- Accuracy;
- Storage limitation;
- Integrity and confidentiality.

An additional principle of accountability states the responsibility of the controller to demonstrate compliance with the rest of the principles. Linking the principles relating to processing of personal data to specific privacy and data protection challenges posed by the AI in 5G networks will give us an understanding of how the AI contributes to dilution of the right privacy and the right to data protection – a potential diminishing of existing privacy and data protection safeguards present in ECHR, CFREU, and EU legislation.

The principle of lawfulness, fairness, and transparency incorporates three requirements and aims to protect personal data from fraudulent gathering and processing. Lawfulness is the first requirement of the principle and means that solutions providers such as AI and ML must respect all applicable laws and regulations concerning processing personal data.

The specific elements of transparency of data processing include: (a) easy access to information regarding processing; (b) clear identity of the controller and the purposes of the processing; (c) provision of information regarding risks, rules, safeguards, and rights. The vendors in AI in 5G networks use cases must ensure compliance with these elements, namely give the information regarding processing personal data, identify the structure of (joint) data controllers and exhaustive explanations regarding purposes of processing and provide the information on possible risks, rules, safeguards and data subject's rights. The complex and dynamic architecture of 5G networks with AI and ML algorithms may prove to be an obstacle to compliance with this principle. Moreover, it is pretty challenging to explain the nature of AI and ML themselves in simple terms. Providing information on specific risks, rules, and safeguards concerning the processing of personal data by AI solutions in 5G networks will be even more difficult. It does not preclude the use of AI in 5G networks. Still, it requires AI solution providers to spend considerable resources on articulating the necessary information to data subjects.

Fairness requires that personal data must be obtained through valid and fair means and not through fraud, deception or without awareness of the data subject (Kuner, Bygrave, Docksey, & Drechsler, 2020). Fairness also implies that controllers and processors should process data only in ways that data subjects would reasonably expect (Information Commissioner's Office, 2020). This essentially means that data subjects must be aware of how their data will be used for AI algorithms enhancing 5G networks. Considering the real-time application of AI in 5G this would not always be possible. Moreover, if the AI or ML algorithm produces erroneous, biased, or discriminatory results due to inaccurate or unreliable data, it may negatively affect the rights of data subjects (Norwegian Data Protection Authority, 2018). Biased or discriminatory results generated by the AI and ML algorithms may substantially affect an individual's right to privacy as stipulated by Article 8 ECHR, which is amplified by a massive number of users in 5G networks. Appropriate safeguards must be implemented to prevent any AI and ML applications from interfering with the right to private life by creating algorithms emphasizing racial or ethnic origin, sexual orientation, political opinion, religion or beliefs, genetic or health status, membership in trade unions. A faulty decision based on a discriminatory algorithm may lead to large-scale consequences in a huge mobile network with many users like 5G.

Purpose limitation is the second processing principle defined by GDPR and requires that personal data is gathered for a specified, explicit, and legitimate purpose. Data controllers AI and ML solutions in 5G networks will need to ensure the provision of free, accessible, specific, and unambiguous information on personal data processing. Such purposes may vary depending on the particular vendor or party processing personal data. Purposes themselves must be clear, precise, explicit and legitimate. Legitimate purposes should not lead to a disproportionate interference with the rights, freedoms, and interests of the data subject favoring the data controller's interests (Kuner, Bygrave, Docksey, & Drechsler, 2020).

The data minimization principle requires limiting the processing of personal data to 'what is necessary in relation to the purposes for which they are processed' providing a clear link with the principle of Purpose limitation. It is vital to ensure that privacy-preserving technologies and techniques accompany the roll-out of virtualized 5G networks. This principle may prove to be extremely challenging to comply with for AI and ML applications in 5G networks. As was stated before, the development of an AI algorithm or model requires an enormous amount of data. The data minimization principle forbids the excessive processing of personal data, which may hinder the development of AI and ML algorithms in 5G networks. Data controllers aiming to use personal data for creating AI-based solutions will have to define specific purposes and characteristics of data (type, amount, level of detail) in order to develop any kind of AI algorithm based on personal data from 5G networks. Re-using old personal data is possible, but data controllers cannot re-use it for any other purpose than the one used to gather this data. Data minimization also requires the data controllers to respect the right to privacy and refrain from using excessively detailed or non-anonymized data (Norwegian Data Protection Authority, 2018). This forces the AI and ML developers to spend considerable resources on carefully defining the purposes of processing and the intended area of application of AI solutions.

The principle of accuracy requires controllers to keep the data up to date and rectify or erase inaccurate and incomplete data (Kuner, Bygrave, Docksey, & Drechsler, 2020). Considering the technological novelties brought by the new generation of mobile networks and complications related to processing vast amounts of data, it is unclear how exactly this principle will be implemented in virtualized 5G networks. In practice, it may be virtually impossible to rectify or erase personal data from the technological perspective. Another technical problem is to keep detailed records of data processing activities using AI and ML technologies and track the flow of personal data. Such records may require substantial computing and storage resources, and their implementation can be complicated.

The storage limitation principle requires controllers to clearly indicate the specific and adequate timeframe for storing personal data. It is prohibited for data controllers to retain gathered personal data beyond the timescale necessary to process data according to the purposes for such processing (Kuner, Bygrave, Docksey, & Drechsler, 2020). Data controllers using AI and ML will need to apply privacy-preserving techniques, like in the case of data minimization principle. While assessing the risks posed by blockchain, Commission Nationale de l'Informatique et des Libertés (CNIL, French Data Protection Authority) stated that some information can be stored without a retention period in exceptional cases.[3] However, in such cases, a data protection impact assessment (DPIA) is required. Certain exceptions are likely possible for AI applications in 5G networks, but they are constrained and must be backed by legal grounds.

Appropriate security (labeled as 'integrity and confidentiality principle') means that processing of personal data is done "in a manner that ensures appropriate security of the personal data, including protection against unauthorised or unlawful processing and against accidental loss, destruction or damage, using appropriate technical or organisational measures" (Information Commissioner's Office, 2020). As mentioned above, multiple actors will be responsible for different elements in 5G networks and may apply different security standards and tools. 5G networks relying on efficiency of AI algorithms may be reconfigured in numerous ways to create various types of networks with different standards and solutions pertaining to privacy and security. AI and ML developers may be wary of sharing their architectural and IT solutions or commercial secrets like business and data practices in order to ensure transparency and react to inquiries from individuals and regulatory bodies. Lack of formal requirements on transparency may result in different objectives in trust pursued by various actors and causes a situation where users cannot obtain reliable information on the security standards and tools used to protect their data. Another concern is related to the use of AI and ML solutions with their unresolved security vulnerabilities. Complex software systems like AI require

3 "If justified by the purpose of the processing and if a data protection impact assessment (DPIA) has proven that the residual risks are acceptable, personal data may exceptionally be stored on the blockchain, in the form of a traditional fingerprint (without a key) or even in cleartext. Indeed, some data controllers may have the legal obligation to make some information public and accessible, without a retention period: in this particular case, the storage of personal data on a public blockchain can be envisaged, provided that the DPIA concludes that the risks for data subjects are minimal". Commission Nationale de l'Informatique et des Libertés. (2018). Solutions for a responsible use of the blockchain in the context of personal data. https://www.cnil.fr/sites/default/files/atoms/files/blockchain_en.pdf.

rigorous testing in a protected environment before any real-life application, but no test can guarantee absolute security. Bugs, deficiencies, and code vulnerabilities may be present and exploited for a considerable time before they are revealed and corrected. The providers of services have to implement state-of-the-art security is solutions. It is hard to identify state-of-the-art in AI and ML, however.

The final principle is the principle of accountability, and it is essentially an obligation to ensure compliance with all other principles listed in Article 5(1) of GDPR. Article 29 WP referred to 'data deluge' effect "where the amount of personal data that exists, is processed and is further transferred continues to grow" (Article 29 Working Party, 2010). The principle of accountability requires data controllers to adopt measures to effectively comply with the data processing principles and demonstrate such compliance. Technological developments contribute to the growth of the amount of data being processed, so the impact of data breaches and data leaks and risk of misuse of personal data increase.[4] Article 29 WP also mentioned legal certainty as one of the challenges towards the implementation of principle of accountability since the requirements of the principle are not detailed (Article 29 Working Party, 2010). The ways and instruments to assess the level and effectiveness of compliance with the principle of accountability may differ and are flexible. According to the Opinion of Article 29 WP, "[f]or larger, more complex and high risk processing, internal and external audits are common verification methods" (Article 29 Working Party, 2010). There is no list of means of verification or authorized methodologies, but the Opinion of Article 29 WP mentions voluntarily certification as one of the instruments of achieving compliance. EU and national legislation usually regulate the accreditation of certification programs. Since 5G networks definitely fall within the category of 'complex and high risk processing' and may contribute to 'data deluge' effect mentioned by Article 29 WP, it is likely that the compliance of vendors and actors processing data through 5G networks with the principle of accountability will be subject to scrutiny by national DPAs.

4 The EDPB maintained that "[…] nature, sensitivity, and volume of personal data increases the risks further, because the number of individuals affected is high, as is the overall quantity of affected personal data […]" and "[a]s usual, during risk assessment the type of the breach and the nature, sensitivity, and volume of personal data affected are to be taken into consideration" linking volume of data to the impact of data breaches. See: EDPB 'Guidelines 01/2021 on Examples regarding Data Breach Notification', 12–19. https://edpb.europa.eu/sites/default/files/consultation/edpb_guidelines_202101_databreachnotificationexamples_v1_en.pdf.

Article 25 of GDPR established principles of data protection by design and by default, which require all controllers to implement technical and organizational measures "[t]aking into account the state of the art, the cost of implementation and the nature, scope, context and purposes of processing as well as the risks of varying likelihood and severity for rights and freedoms of natural persons posed by the processing" in order to protect the rights of data subjects both at the time of the determination of means for processing and during the processing itself. Therefore, the implementation of technical and organization measures is needed to ensure that 'by default' only personal data which is necessary to achieve purposes according to purpose limitation is processed.

The principle of data protection by design and by default requires developers of applications, services and products to consider privacy and data protection principles early at the development of their respective goods and services. The idea is that incorporation of privacy and data protection requirements early substantially improves the privacy and data protection resilience and compliance of applications, services and products (Kuner, Bygrave, Docksey, & Drechsler, 2020). These legal provisions are aimed at ensuring the high efficiency of protection of data subjects' rights. The data controllers have a legal duty to implement technical and organizational measures to ensure compliance with data protection principles. These measures go beyond the security and reliability of software and include business practices, organizational principles etc. (Kuner, Bygrave, Docksey, & Drechsler, 2020). There is no exhaustive list of such measures, but they include, among others, training of personnel, automated erasure of outdated data, economic approaches to processing data, evading excessive processing of data or storing unnecessary data. AI and ML providers also must adopt a risk management approach to implement appropriate and effective technical and organizational measures to comply with the principle of data protection by design. The aim is to ensure the protection of data subjects' right to privacy and protection of personal data.

Objectively, compliance with this principle may prove to be a major obstacle for implementation of AI algorithms in 5G networks. The path to compliance with data protection by design principle is mainly linked with principles relating to processing of personal data, specifically data minimization and storage limitation. As was stated before, the complex and dynamic architecture of both AI and 5G, involvement of multiple actors from multiple jurisdictions, possible hesitation of involved actors to comply with transparency principle or give up commercially valuable personal data to comply with data minimization and storage limitation principles are factors hindering the fulfillment of legal requirements and ensuring the protection of data subjects' rights. This inevitably leads to tougher government control and loss of trust in the ability of technology to protect privacy.

Lastly, the implementation of AI and ML solutions in 5G will almost certainly require performing the Data Protection Impact Assessment (DPIA) which is an important instrument and requirement devised in the GDPR to get a clear understanding of possible impact of using certain technology or setting for processing personal data. According to Article 35(1) of GDPR, "[w]here a type of processing in particular using new technologies, and taking into account the nature, scope, context and purposes of the processing, is likely to result in a high risk to the rights and freedoms of natural persons, the controller shall, prior to the processing, carry out an assessment of the impact of the envisaged processing operations on the protection of personal data". Criteria required for such assessment are evident concerning AI in 5G networks:

- New technologies – both AI and 5G are disruptive novel technologies;
- Nature and scope – AI and ML will be used at a large scale up to becoming omnipresent in 5G networks and will continuously process personal data;
- High risk – due to large amounts of data and unclear safeguards the AI and ML solutions in 5G networks can be qualified as high risk.

DPIA is crucial but resource-consuming procedure. It is possible that many solution providers in 5G, including but not limited to AI, will hesitate to gather and provide full and accurate information in order to perform the DPIA.

4 Conclusion

It is evident, that numerous challenges related to implementing AI and ML solutions in 5G networks may impact the privacy and data protection safeguards and lead to the dilution of the right to privacy and diminishing the role of consent to personal data protection.

The telecommunications industry relies on technological novelties such as AI and ML to deliver the better, faster, and more reliable connections to users. The personal data processed via 5G networks may prove to have a high commercial value, attracting a considerable number of vendors and application developers. However, the new generation of mobile networks is vulnerable to software bugs, data leaks, lack of coordination between different parties involved in processing personal data, and cyber attacks. Other issues include the dynamic and complex architecture of 5G networks, issues of sharing confidential data on software elements or business practices, and resource-demanding compliance procedures.

Implementation of effective privacy-preserving technologies, data protection measures, and security standards is a requirement of law and a necessity enabling the operation of 5G network as a whole. As was noted by S. Rizou, E. Alexandropoulou-Egyptiadou, and K. E. Psannis (2020), "[t]he scope of privacy protection would be not only the effort to avoid the administrative fines of millions of euro, but to establish from the beginning of 5G technology, a fair integrated treatment for data protection rights". Adherence to GDPR requirements, namely to principles relating to processing personal data, the principle of data protection by design and by default is crucial for ensuring privacy and data protection. Users must be able to easily obtain full and clear information on the processing of personal data, the data controllers, purposes of the processing, profiling, and automated decision-making, and the ways to exercise their rights to rectification and erasure of data. AI and ML providers must ensure that the consent to processing personal data is given freely, and it is specific, informed, and unambiguous. Considering the use of edge computing and cloud services and the option of distributed processing of personal data in 5G networks, traceability mechanisms must be implemented, which may prove to be a substantial challenge (AEPD, 2020). The fact that 5G contributes to generating an immense amount of personal data highlights the importance of effective data minimization measures.

The privacy, security, and data protection in 5G networks largely depend on the 5G value chain actors. Ultimately it is up to AI and ML providers to comply with the EU privacy and data protection legal framework. Their decisions, management practices, procedures applied will have a decisive impact on the functioning of 5G networks in general.

Acknowledgement

This research has received funding under EU H2020 MSCA-ITN action 5GhOSTS, grant agreement no. 814035.

Bibliography

AEPD. (2020). *Introduction to 5G technologies and their risks in terms of privacy.* Agencia Española de Protección de Datos.

Article 29 Working Party. (2010). *Opinion 3/2010 on the principle of accountability, 00062/10/EN WP 173.*

Article 29 Working Party. (2011). *Opinion 13/2011 on Geolocation services on smart mobile devices, 881/11/EN WP 185.*

BEREC. (2019). *A first assessment based on stakeholder inputs – Report on the impact of 5G on regulation and the role of regulation in enabling the 5G ecosystem.*

Blackman, C. and Forge, S. (2016). *European Leadership in 5G.* Directorate General for Internal Policies of the European Parliament.

European Commission. (2020). Communication from the Commission to the European Parliament and the Council *Data protection rules as a pillar of citizens empowerment and EUs approach to digital transition – two years of application of the General Data Protection Regulation.* https://eur-lex.europa.eu/legal-content/EN/TXT/?uri=CELEX%3A52020SC0115.

European Court of Human Rights. (2020). *Guide on Article 8 of the European Convention on Human Rights.* Council of Europe.

Haider, N., Zeeshan, M. and Imran, M. (2020). Artificial Intelligence and Machine Learning in 5G Network Security: Opportunities, advantages, and future research trends.

Holvast, J. (2009). History of Privacy. In F. H. S. Matyáš V., *The Future of Identity in the Information Society. Privacy and Identity. IFIP Advances in Information and Communication Technology.* Berlin, Heidelberg: Springer.

IEEE Future Networks World Forum (2021). AI and Machine Learning Track, https://ieee-wf-5g.org/ai-ml-track/.

Information Commissioner's Office. (2020). *Guide to the General Data Protection Regulation (GDPR).* ICO.

Kosta, E. (2013). *Consent in European Data Protection Law (Nijhoff Studies in European Union Law).* Martinus Nijhoff.

Kuner, C., Bygrave, L. A., Docksey, C., and Drechsler, L. (2020). *The EU General Data Protection Regulation (GDPR): A Commentary.* Oxford: Oxford University Press.

Madhusanka, L., Salo, J., Braeken, A., Kumar, T., Seneviratne, S., & Ylianttila, M. (2018). 5G Privacy: Scenarios and Solutions. *2018 IEEE 5G World Forum (5GWF).*

Morocho Cayamcela, M., & Lim, W. (2018). Artificial Intelligence in 5G Technology: A Survey. *International Conference on Information and Communication Technology Convergence (ICTC).*

Norwegian Data Protection Authority. (2018). *Artificial intelligence and privacy.* Norwegian DPA.

O'Dea, S. (2020). *Mobile technology share by generation 2016–2025.* Statista.

Parliamentary Assembly. (2017). *Technological convergence, artificial intelligence and human rights.* Council of Europe.

Rizou, S., Alexandropoulou-Egyptiadou, E., & Psannis, K. E. (2020). GDPR Interference With Next Generation 5G and IoT Networks. *IEEE Access*, 108052–108061.

Warren, S., & Brandeis, L. (1890). The Right to Privacy. *Harvard Law Review*, 193–220.

Zhang, Y. (2018). *Network Function Virtualization.* Hoboken, NJ: Wileys, IEEE Press.

CHAPTER 6

The Empowerment of Platform Users to Tackle Digital Disinformation: the Example of the French Legislation

Irène Couzigou

Disinformation is not a new phenomenon. What is new, however, are the scale and delivery methods of disinformation on the Internet. Digital information is relayed quickly, in minutes, and massively, to thousands of people. In addition, the diffusion of digital information can be amplified through the resort to technologies. Thus, automated or semi-automated actors – 'bots' – can spread erroneous content through biased retweets and likes. Trolls, individuals or groups, can saturate websites with false data (Vilmer et al., 2018: 83–84).

This chapter understands digital disinformation as content that is knowingly false and disseminated online to intentionally deceive the public.[1] Digital disinformation can have negative consequences on public policies. Hence, public health disinformation affected the fight against the COVID-19 pandemic (Butcher, 2021: 7). Disinformation may also undermine the trust of citizens in democratic institutions. That would be the case if citizens received false information about the candidates who are running in an election or about a proposition submitted to referendum. For instance, digital disinformation likely to have been initiated by Russia and aimed at swaying voting behaviour was spread in electoral campaigns preceding the election of the American President in 2016, the referendum on the EU membership of the United Kingdom in 2016, and the referendum on the independence of Catalonia in 2017 (Vilmer et al., 2018: 49).

The French electoral campaign leading up to the election of the president of the republic in 2017 was also subject to digital disinformation. The purpose of that disinformation was to harm the reputation of candidates, especially that of Emmanuel Macron who was a front runner. In particular, just before the end of the official campaign, hacked e-mails from Macron's close collaborators were posted online, together with fake documents.[2] A link to

1 On the contrary, misinformation is defined as inaccurate content not out of malicious intent but merely error or ignorance.

2 The uploaded content included 9.2 GB of compressed data, including 21,075 emails.

 | DOI:10.1163/9789004547261_008

the Twitter thread with the hashtag #MacronLeaks was rapidly posted and massively shared on Twitter by automated accounts and real people (Mohan, 2017). Although the disinformation strategies did not significantly influence French voters and prevent Emmanuel Macron from winning the presidential election, President Macron vowed to introduce a law that specifically addresses false information disseminated online during national electoral periods (Macron, 2018).

The ordinary law on the fight against the manipulation of information was passed at the end of 2018, following a difficult legislative journey (Law no. 2018–1202). The organic law on the fight against the manipulation of information was adopted on the same day.[3] It includes the provisions of the ordinary law into the law on the election of the President of the Republic (Organic law no. 2018–2101). These are laws aimed to reinforce the honesty of electoral debate and to avoid citizens being influenced by digital disinformation when exercising their vote. Their implementation is limited to general elections of members of the National Assembly, of members of the Senate, of the president of the republic, of the French representatives in the European Parliament, and national referenda which, due to their national significance, are more likely to be targeted by disinformation campaigns than local elections.

The ordinary and organic laws against information manipulation were submitted to the Constitutional Council, the supreme constitutional court in France. The Constitutional Council declared those laws compatible with the Constitution subject to certain reservations relating to their interpretation (Constitutional Council, 2018–773, Art. 1 and 2; Constitutional Council, 2018–774, Art. 1 and 2). In particular, the Constitutional Council limited the scope of the new legislation in order to respect the right to freedom of expression and communication. Indeed, this right applies to all communications, whether conducted in the physical or cyber world.[4] Further, the right to freedom of expression and communication is guaranteed by the Declaration of the Rights of Man and the Citizen of 1789 (Declaration of the Rights of Man and the Citizen, 1789, Art. 11), and as such enjoys constitutional value under French law (Constitutional Council, 1971, para. 2). It is also enshrined in the European

3 Under French law, an organic law completes a provision of the French Constitution but without being included in the Constitution; it thus enjoys a higher value than an ordinary law in the French hierarchy of norms.

4 States have asserted their sovereign authority and jurisdiction over cyber activities conducted on their territory, and thus the implementation of national and international norms deriving from the principle of sovereignty, including human rights law.

Convention on Human Rights (1950, Art. 10) and the International Covenant on Civil and Political Rights (1966, Art. 19) to which France is a party. The right to freedom of expression may be limited as 'prescribed by law' (European Convention on Human Rights, 1950, Art. 10). Thus, the French legislature is entitled 'to institute provisions to bring an end to the abuse of the right to exercise freedom of expression and communication which infringes on public order and the rights of others' (Constitutional Council, 2018–773, para. 14). Limits to the right to freedom of expression and communication must however be as narrow as possible and be necessary and proportionate to the objective sought (Constitutional Council, 2018–773, para. 15).

The laws against the manipulation of information are for the moment the sole European legislation that specifically tackle digital disinformation that intends to deceive the public. This chapter assesses the different means adopted by the new French legislation and draws lessons for the preparations of other legislations directed against the spread of misleading false information online. The EU, in particular, is preparing a Digital Services Act that should include provisions relating to digital disinformation. The ordinary law of 22 December 2018 is the main arsenal addressing false information. The organic law simply refers to it. This article will thus focus on the ordinary law relating to the fight against the manipulation of information, in the light of its interpretation by the Constitutional Council. When not otherwise specified, references to the law against the manipulation of information are references to the ordinary law.

This chapter will examine what role that online platforms that transmit or store digital content can play in the fight against digital disinformation. It will demonstrate that while online platforms can reduce the impact of digital disinformation, their role should remain minor in the reaction to the spread of that disinformation. This article will then study the powers attributed to the French audiovisual regulatory authority to prevent, suspend, or react to the diffusion of digital disinformation and demonstrate that those powers are constrained by the need to respect the right to freedom of expression, subsequently moving to the attributions of the judge sitting for urgent matters created by the new legislation to tackle disinformation disseminated online. As will be seen, her/his attributions are also limited by the right to freedom of expression. Next, the state of information and media literacy in France will be addressed. Finally, this chapter will conclude that regulating the diffusion of disinformation online is often contrary to the right to freedom of expression. It will argue that the fight against digital disinformation should focus on preventing the consequences of that disinformation.

1 A Balanced Role of Online Platforms in Preventing the Impact of, and Reacting to, Digital Disinformation

A means to address the problem of digital false information that aims to impact electoral results is to prevent that impact. Such could be the case if Internet users are made aware of the origin, and support received by, that information that is the most likely to affect their voting behaviour. For instance, if Internet users knew that a piece of information directed against a political party has been sponsored by another party, they may apprehend that information with a critical mind and be less influenced by it. Some online platforms, conscious that diffusing false content affects their reputation, had already taken proactive measures to increase the transparency and reliability of the content they transmit. Thus, shortly before the 2017 French presidential election, Facebook suppressed fake accounts in France and eliminated the sponsoring of 'clickbaits', links with attractive titles that refer to other websites (Frassa, 2018: 13). Furthermore, Google partnered with over 30 media outlets to create the CrossCheck fact-checking platform (Brattberg and Maurer, 2018). CrossCheck intended to report false, misleading, and confusing content that spread online during the French presidential campaign and thereby to warn the public about that content.

The law against manipulation of information imposes new information transparency obligations on online platforms that have a broad audience and may, as such, influence opinion. Thus, only large-scale online platforms fall under the purview of the new legislation: they must be covered by Article 111–7 of the Consumer Code and have an activity exceeding five million connections per month on French territory (Decree no. 2019–297).[5] Those platforms exercise a pallet of diverse activities: they provide social media (for instance Facebook or Instagram), web portals (Yahoo Portal or Google), search engines (Google or Yahoo Search), marketplaces (Amazon or Airbnb), or content sharing (YouTube or Dailymotion).

The law of 22 December 2018 differentiates between information transparency requirements, specific to electoral campaigns and others, applicable permanently. Furthermore, the new legislation asks online platforms to react to

5 According to Art. 111–7 of the Consumption Code, an online platform operator is "any natural or legal person offering, on a professional basis, whether remunerated or not, an online communication service to the public based on: (1) Classification or referencing, by means of computer algorithms, of content, goods or services offered or put online by third parties; (2) Or bringing together several parties for the sale of a good, the provision of a service or the exchange or sharing of content, a good or a service".

false information they are aware of. Important online platforms have to respect strict transparency requirements during the three months preceding the first day of the month when general elections or referenda are held and up to the date of those elections or referenda. In accordance with those requirements, large online platforms must provide their users with fair, clear, and transparent information about the identity of the natural person or about the company name, registered office, and corporate object of the legal person and on whose behalf it has declared to act, when that natural or legal person pays the platform for promoting information 'related to a debate of general interest' (Law no. 2018–1202, Art. 1). The same platforms must also provide their users with fair, clear, and transparent information about the use of their personal data when they promote information 'related to a debate of general interest' (Law no. 2018–1202, Art. 1). The amount of remuneration received by online platforms covered by the law of 22 December 2018 in return for promoting information of general interest must also be made public, when this amount exceeds E100 (Law no. 2018–1202, Art. 1; Decree no. 2019–297, Art. 1). The Constitutional Council limited the information of general interest subjected to the transparency requirements imposed by the new legislation to that information related to an electoral campaign (Constitutional Council, Decision no. 2018–773 DC, 20 December 2018, para. 8).

Thus, the new legislation provides users of large online platforms with the means to assess the degree of reliability of information linked to an electoral campaign when this information is disseminated against substantial payment. The law against manipulation of information is intended to address the questionable financing of websites that occurred, for instance, during the 2016 American presidential election (Vilmer et al., 2018: 144). It does not prohibit sponsorship of information in electoral periods but frames it through a transparency framework. All information to be given by online platforms by virtue of the law of 22 December 2018 must be aggregated in a register made available to the public by electronic means (Law no. 2018–1202, Art. 1).

The law against manipulation of information also introduces transparency requirements that must be permanently respected by large-scale online platforms, during and outside electoral periods. French law had already imposed general transparency obligations on important digital platforms. Thus, users of large platforms must have access to the platforms' terms of service as well as to the classification, referencing, and dereferencing procedures of the content they disseminate. Further, users of important platforms should be made aware of what content is sponsored (Consumer Code, Art. L. 111–7).

The law of 22 December 2018 reinforces transparency obligations for important online platforms, particularly in relation to information likely to affect electoral processes. The new law requires large online platforms to cooperate in the fight against false information that may compromise the integrity of upcoming votes or, alternatively, disrupt public order. False information must be understood as including objectively incorrect or misleading allegations or accusations (Constitutional Council, 2018–773, para. 86). First, platform operators must set up an easily accessible and visible mechanism for users to report false information likely to affect public order or the honesty of one of the elections or referenda to which the law of 22 December 2018 applies (Law no. 2018–1202, Art. 11 para. 1). Second, platform operators must adopt a range of measures to combat false information that may alter public order or the honesty of general elections or referenda. Here, the law of 22 December 2018 does not impose precise transparency obligations on online platforms – contrary to the transparency obligations imposed during electoral periods – but a duty to cooperate in the combat against information manipulation. Digital platforms enjoy a broad margin of appreciation as to the measures they can take (Law no. 2018–1202, Art. 11 para. 1). Those measures may relate to the transparency of algorithms that sort, reference, and select content and detect false information; the promotion of reliable content; the fight against accounts that disseminate false information through the suppression of those accounts' content or the blocking of those accounts; the information of users about the identity of the natural person or entity who remunerates a platform for promoting content of general interest; the information of users about the nature, origin, and modalities of dissemination of content of general interest; and the support of media and information literacy (Law no. 2018–1202, Article 11 para 1).

The new law against information manipulation recognises that online platforms are more than simple transmitters of information. Internet platforms do not just show information randomly – they sort, select and rank news and information via their algorithms. Platform companies are therefore asked to cooperate in the fight against false information. The new legislation cannot go further and impose on online platforms an obligation to police the information they store or transmit and then to report false information to public authorities or to take down that information themselves. Indeed, in accordance with the law for confidence in digital economy that transposes the European Union directive of 8 June 2000 on electronic commerce into French law, online platforms are not liable for the content they disseminate provided they act expeditiously to remove or disable access to unlawful content when they obtain actual knowledge of such content (Law no. 2004–575).

More generally, requiring from online platforms to search for, and to suppress, false information likely to affect the integrity of elections or referenda, may incite platforms to over-remove information for fear of penalties (Report,

2016, Para. 52). This may be especially the case as false information within the purview of the law of 22 December 2018 cannot be clearly circumscribed. If platforms were obliged to moderate the content they store or diffuse, they may suppress harmless information. This would undermine the right to freedom of expression of their users, whose limits must be necessary and proportionate.

Digital platforms covered by the law against information manipulation must designate a legal representative who will act as referent interlocutor on the French territory for the application of that law (Law no. 2018–1202, Art. 13). Compliance by online platform operators with their new transparency obligations is entrusted to the audiovisual regulatory authority, the Superior Audiovisual Council. This Council enjoys legal personality and decisive powers exercised independently from the French executive (Law no. 2013–1028, Art. 33). Online platform operators covered by the law of 22 December 2018 must submit annual reports to the Council explaining how they implemented that law and the problems they came across (Law no. 2018–1202, Art. 11 para. 1). The Superior Audiovisual Council may publish recommendations and reports relating to the implementation of the law against information manipulation by digital platforms (Law no. 2018–1202, Art. 12). In its first report, the Council stressed the importance of accuracy, clarity, fairness, and transparency in the fight against false information. In particular, it recommended to platform operators that carry out reviews of content reported as false information to adopt standards and reference sources. It also encouraged the creation of partnerships between platforms and fact-checkers (Superior Audiovisual Council, 2020).

In monitoring whether and how digital platforms tackle false information based on the law of 22 December 2018, the Council should assess whether online platforms act within the limits of freedom of expression. Indeed, whereas online platforms, as non-State actors, are not legally bound by international human rights law, they have nevertheless a corporate social responsibility to comply with it (Report, 2011, Principle 11). In conformity with the law of 22 December 2018, 10 online platforms have created a reporting mechanism for false information. As they indicated to the Superior Audiovisual Council, those platforms addressed the false information that was reported to them through diverse measures, ranging from reducing the visibility of disinformation to removing it, or even taking down accounts disseminating disinformation (Superior Audiovisual Council, 2020: 24–25).

It remains to be seen in the long term how often platform users will report false information and how platforms will address the reported information. When removing false information or online accounts, online platforms should be guided by necessity and proportionality principles (Report, 2018, para. 47).

They should thus limit the removal of content to the most obvious cases of distorting false information. That information can be flagged by users or the platforms themselves. Given the risk of automated suppression of legitimate content by algorithms, a human reviewer should ultimately make the decision about whether it should be suppressed or not (Dias, 2020: 610). Further, platforms should notify the relevant users why content was taken down, or accounts suspended or terminated, and allow them to appeal to an independent body (Association for Progressive Communications, 2018: 16). If online platforms do not conform with the right to freedom of expression when moderating content, the French State should compel them to do so. Indeed, States have a positive obligation to create a regulatory environment in which all users' rights can be respected (Report, 2011, Principle 3). States cannot use Internet intermediaries to implement public policy – the fight against certain digital information – and *de facto* escape obligations they would have if they conducted that policy themselves (Land, 2019: 303).

2 A Constrained Action of the Audiovisual Regulatory Authority in Preventing and Reacting to Digital Disinformation

The law against manipulation of information establishes specific tools to counter digital disinformation disseminated under the control or influence of a foreign State with the purpose of influencing electoral results in the French State. The new legislation reacts to the circulation of disinformation directed against Emmanuel Macron during the presidential campaign by the Russian State-funded news outlets, RT, formerly called *Russia Today*, and *Sputnik* (Vilmer et al., 2018: 144). As denounced by Emmanuel Macron himself, once President of the Republic, 'Russia Today and Sputnik were organs of influence during this campaign that repeatedly produced untruths about me and my campaign' (Jambot, 2017).

Broadcasting by radio or television services must be authorised through an agreement concluded with the Superior Audiovisual Council if they are distributed by networks that do not use the frequencies assigned by the Council and if their annual budget exceeds a certain amount.[6] The law against information manipulation gives to the Superior Audiovisual Council the power to suspend, interrupt, or prevent authorisation to broadcast to radio or television channels controlled by, or under the influence of, a foreign State when those

6 EUR 150,000 for a television service; EUR 75,000 for a radio service.

channels deliberately disseminate false information likely to affect the honesty of votes in general elections or referenda. In order to ensure the compatibility of the law of 22 December 2018 with the right to freedom of expression, the Constitutional Council limited the scope of "false information" that justifies a suspension or termination of an authorisation to broadcast. False information must then be understood as including incorrect or misleading allegations or accusations that are objectively and completely false – not just an opinion or parody (Constitutional Council, 2018–773, para. 51 and 61).

First, the Superior Audiovisual Council can suspend the authorisation to broadcast to a radio or television service within the three months preceding the occurrence of a general election or referendum when this service diffuses false information that is very likely to influence voters (Law no. 2018–1202, Art. 6). With this new provision, the French legislature intends to prevent citizens from being deceived when exercising their vote by disinformation spread with the control or support of a foreign State.

Second, the Superior Audiovisual Council can terminate an agreement authorising the broadcast of a radio or television service and concluded with a legal person controlled by or placed under the influence of a foreign State. The Council can do so when the radio or television service harms the fundamental interests of the nation, including the regular functioning of its institutions, especially by disseminating false information before general elections or referenda. The 'fundamental interests of the Nation' allude to a concept already defined under French law (Criminal Code, Art. L. 401–1; Homeland Security Code, L. 811–3).[7] In order to assess whether a radio or television service infringes the fundamental interests of the nation, the Superior Audiovisual Council can take into consideration the content that the company that entered into agreement with the Council or the legal person controlling it has broadcasted on other electronic communication services (Law no. 2018–1202, Art. 8).

Finally, the Council can refuse to conclude an agreement to broadcast by a radio or television channel, directly or indirectly controlled by a foreign State, if the broadcast involves a serious risk of infringing the fundamental interests of the nation, in particular the regular functioning of its institutions (Law no. 2018–1202, Art. 5). To demonstrate the risk of fundamental interests being

7 Art. L. 401–1 Criminal Code States: "The fundamental interests of the nation are understood within the meaning of this title of its independence, the integrity of its territory, its security, the republican form of its institutions, the means of its defense and its diplomacy, the safeguarding of its population in France and abroad, the balance of its natural environment and its environment and essential elements of its scientific and economic potential and its cultural heritage."

infringed, the Council can refer to, among others, the content diffused on other electronic communication services by the company asking for authorisation to broadcast or by the legal person controlling it (Law no. 2018–1202, Art. 5). Therefore, the Superior Audiovisual Council could refuse authorisation to broadcast to a company if that company disseminated false information with the purpose of influencing voting behaviour and thereby the functioning of the institutions of another State.

When assessing whether a radio or television channel is controlled by a State, the Superior Audiovisual Council should refer to the notion of 'control' within the meaning of the French Commercial Code (Law no. 2018–1202, Art. 5, 6, and 8). In conformity with this Code, a legal person controls a company when it has most of the voting rights in the general meetings of this company and thus determines the decisions taken at those meetings (Commercial Code, Art. L. 233–3). The Council could also apply the concept of "control" within the meaning of international law. For Article 8 of the Draft Articles on Responsibility of States for Internationally Wrongful Acts that codifies a customary provision: "[t]he conduct of a person or group of persons shall be considered an act of a State under international law if the person or groups of persons is in fact acting under the direction or control of, that State in carrying out the conduct" (Draft Articles, 2001, Art. 8). The degree of control for the conduct of a (physical or legal) person to be attributable to a State must be high (Draft Articles, 2001, Art. 8). Therefore, under both French or international law, a radio or television service should be seen as being controlled by a State and acting on behalf of that State when it is financed by that State and its activities are decided by it. The law of 22 December 2018 does not define what is an audiovisual service "placed under the influence" of a foreign State. Being influenced by a State implies less closeness to that State than being controlled by it. Thus, a radio or television service supported or encouraged by a State should be regarded as under the influence of that State.

In conclusion, the Superior Audiovisual Council enjoys a broad discretion in determining whether a radio or television service is under the control or influence of a State. The powers of the Council in the suspension or interruption of false information that is diffused by such service and that is likely to affect the outcome of general elections or referenda are however limited by the right to freedom of expression. Indeed, the Superior Audiovisual Council can act against this radio or television service only if the information it transmits is incorrect, misleading, or objectively false. Furthermore, the new powers of the Superior Audiovisual Council are also politically limited. Indeed, the Council may hesitate to act against a channel controlled or influenced by a foreign State because such action may trigger a counter-reaction of the foreign

State. Another means to tackle digital false information adopted by the law of 22 December 2018 is the creation of a specific judicial procedure.

3 Limited Judiciary Power Reacting to Digital Disinformation

The legislature has established new legal proceedings to address content spread online by creating a judge sitting for urgent matters who may prescribe any measure to stop the dissemination of inaccurate or misleading allegations or imputations of facts likely to alter the honesty of elections (Law no. 2018–1202). This judge can, for instance, order an online platform to suspend or suppress content or close a user's account that spreads false information (Dreyer, 2019: 33). The interlocutory proceedings do not target the author (often unknown) of false information, but the Internet access provider or Internet content host. Measures ordered by the judge sitting for urgent matters must be necessary and proportionate to their objective – to bring an end to false information being diffused – with minimum impact on the right to freedom of expression (Constitutional Council, 2018–773, para. 25).

The referral to the new emergency judge is broad: it can be made by the public prosecutor, any candidate, any party or political group, or any person who considers himself victim of false information (Law no. 2018–1202, Art. 1). A case for manipulated information can be brought before the judge only during the three months preceding the first day of the month of general elections or referenda and until the date of the last round of those elections or referenda (Law no. 2018–1202, Art. 1). The legislature restricts the implementation of the new interlocutory proceedings to the periods when information manipulation is likely to have the highest impact on voting behaviour.

Once a case has been referred, the judge sitting for urgent matters has only 48 hours to rule (Law no. 2018–1202, Art. 1). Given the speed at which information spreads online, the emergency judge needs to decide quickly on a referral for digital disinformation. On the other hand, however, the period of 48 hours can be too brief to find out whether the information is false and impacts on the integrity of an election. Evaluating whether certain content is inaccurate or misleading is particularly difficult in an electoral campaign when many people may express opinions that could be regarded as erroneous. Further, assessing whether certain content may distort the honesty of an election before that election has even taken place is complex (Guillaume, 2019: 5). There is a risk that the judge for urgent matters would censure digital information that was not false and/or not meant to influence the behaviour of voters (La rédaction, 2019). As will be seen below, however, the judge for urgent matters can act only against apparent false information that has a manifest risk of affecting the behaviour of electors (Constitutional Council, 2018–773, para. 16). The judge should be able to detect that obvious disinformation within 48 hours.

Mindful of the need to respect the right to freedom of expression, the Constitutional Council narrowed down the scope of the notion of 'false information': 'These allegations or accusations ... are those for which it is possible to objectively demonstrate falseness' (Constitutional Council, 2018–773, para. 21). Furthermore, 'the allegations or accusations in question can only justify such a measure [based on the interlocutory proceedings] if the incorrect or misleading nature is apparent' (Constitutional Council, 2018–773, para. 23). The risk of alteration by those allegations or accusations of the honesty of elections 'must also be apparent' (Constitutional Council, 2018–773, para. 23). Moreover, inaccurate or misleading allegations or imputations are covered by the law of 22 December 2018 only if their dissemination meets three conditions: it must be deliberate – the false information is disseminated on purpose; artificial or automated – the false information is sponsored or spread through bots; and massively transmitted by an online communication service – the false information is seen by a high number of a platforms' users (Law no. 2018–1202, Art. 1). The introduction of these requirements to be fulfilled by the dissemination of false information was intended to guarantee that the new proceedings apply only to false information aiming to manipulate the public. Indeed, the new legislation does not want to target false information, but manipulation by false information (Moutchou, 2018: 29). In a democracy, Internet users should be able to share information, whether true or false. Freedom of expression is especially important during electoral campaigns (Court of Cassation, Correctional chamber, 2015; Constitutional Council, 2018–773).

As the scope of the new interlocutory proceedings is very narrow, one can question their usefulness. Thus, the new legal injunction was implemented only once. The competent Tribunal for the new emergency legal action, the Paris 'Tribunal of big instance', was invoked during the European elections of May 2019, by two politicians who intended to demonstrate its uselessness.[8] They asked the Tribunal to order SAS Twitter France to withdraw a tweet published by the Twitter account @CCastaner on May 1, 2019. In this tweet, Christophe Castaner, then minister of the interior, wrote that the Pitié-Salpêtrière hospital had been attacked on the margins of the May Day demonstration (Mounier, 2019). The 'Tribunal of big instance' held that '[i]t resorts that if the statement

8 The "Tribunal of big instance" is a tribunal of general jurisdiction that hears cases at first instance when the claim is more than EUR 10,000 and cases which are not specifically allocated to specialist courts.

drafted by Mr. Christophe Castaner appears exaggerated … this exaggeration relates to facts that, themselves, are real …' Thus, 'the condition that the allegation must be manifestly inaccurate or misleading is not met' and the Tribunal declared the request inadmissible (Judgment of the Paris 'Tribunal of big instance', 2019).

In conclusion, limited by the right to freedom of expression, the new legal injunction introduced by the law against information manipulation can order the suspension or suppression of only limited digital content. Another tool chosen by the new legislation in the fight against manipulation information is the increase of media and information literacy.

4 The Need for Media and Information Literacy to Prevent the Impact of Disinformation

The law of 22 December 2018 provides for strengthening media and information literacy in schools, especially for content disseminated over the Internet and as part of moral and civic education (Law no. 2018–1202, Art. 16). The new legislation intends to help children to become responsible digital citizens, able to develop a critical mind in the use of the Internet. In France, media and information literacy should be offered at all levels of education. Cycle 2, from children aged six to eight years, is dedicated to fundamental learning, while cycle 3, for children from nine to twelve years, is devoted to their consolidation. Students in cycle 3 should be able to interrogate information sources and the reliability of those sources. They should be capable of distinguishing trustworthy information from opinions, rumours, or propaganda. Children should also be able to recognize inappropriate behaviour and content, such as an attempt at manipulation or hateful content. Finally, in cycle 4, for children from 13 to 15 years, media and information literacy should be educated through all classes, be it French, history-geography, physics-chemistry, life and earth sciences, or foreign languages (Education Code, Art. L. 321–3; Law no. 2018–1202, Art. 17).

The training of teachers and other education personnel must also involve the development of media and information literacy skills (Law no. 2018–1202, Art. 18). In practice, however, media and information literacy does still not belong to the curriculum in many French schools. Indeed, only a few teachers are trained in new information technologies. External speakers must often be invited in classes to raise awareness about information manipulation (Studer, 2018: 21). Following the increase of the culture ministry's budget for media and information literacy in 2018, funds were invested in the support of civil society actors (e.g., associations and journalists) in educating children on media literacy (Bancaud, 2019). In addition, the Superior Audiovisual Council recommended to online platform operators to support projects and establish partnerships that contribute to media literacy, information literacy, and education on digital tools (Superior Audiovisual Council, 2019: 5; 2020: 69).

Education in new information and communication technologies is an essential way of preventing or at least diminishing the influence exercised by false information on Internet users. If Internet users were better educated into the resort to digital environment, they would be able to critically approach, assess and filter information and be more resilient against information manipulation, including political manipulation (Communication from the Commission, 2020: 10). If citizens knew the codes and languages of digital media, they would be more able to distinguish quality information from false information. Digital skills are all the more crucial given that biased and polarising content is often sophisticated and hard for verification systems to detect. The group of experts established by the European Commission to advise on policy initiatives in the fight against fake news and disinformation spread online also stressed the importance of media and information literacy (European Commission High-Level Group of Experts, 2018: 25–27).

It is important to educate people of all ages about the way technology systems work. Media and information literacy should be offered to children, students and prospective teachers, and also to adults outside training, with a particular focus on professionals who are more likely to react and thereby limit the impact of digital disinformation. For instance, the National Cybersecurity Agency organised a workshop on cybersecurity for French political parties at the beginning of the last presidential campaign to alert them of the risk of harmful cyberoperations. This training contributed to the quick and efficient reaction of Macron's political party towards the Macron leaks (Vilmeret al., 2018: 113–114).

5 Conclusion

The new legislation of 22 December 2018 adopts a range of tools to tackle digital disinformation aimed at influencing voting behaviour. First, the legislation imposes new transparency obligations on large-scale online platforms. In the three months preceding general elections and referenda, important digital platforms have to inform their users about the identity of natural or legal persons who sponsor information related to electoral campaigns. They must also be transparent about the remuneration they receive when it reaches a certain amount. Through those measures, it is hoped that platforms' users will be able to evaluate the reliability of sponsored information and not be affected by it when participating in a general election or referendum. The law of 22 December 2018 also asks digital platforms to take diverse measures to increase the transparency of information, especially that likely to alter the honesty of general elections or referenda. While the law of 22 December 2018 accepts that online platforms may suppress certain false information they come across – e.g. through reporting by users –, it cannot impose on digital platforms an obligation to proactively look for false information and to block or take down false information. This would indeed be contrary to the e-commerce directive. Further, this would incite platforms to over-remove content and thereby impact on the right to freedom of expression of their users.

Second, the new legislation intends to prevent, or react to, the diffusion of disinformation by radio and TV channels directly or indirectly controlled by a foreign State. The law against information manipulation provides to the Superior Audiovisual Council decisive powers to suspend, terminate, or reject an authorisation for broadcasting to a radio or television service controlled by, or placed under the influence of, a foreign State when that service disseminates false information, particularly that likely to affect electoral results. To respect the right to freedom of expression, the Constitutional Council has however limited the scope of false information that justifies a suspension or interruption of broadcasting to the sole information for which it is objectively possible to demonstrate falseness.

Third, the new legislation establishes interlocutory proceedings against digital false information. A new judge can within 48 hours order the suspension or suppression of digital content likely to impact on voting behaviour. Again, due to the right to freedom of expression, the Constitutional Council limited the implementation of the new interlocutory proceedings to only a few information. Thus, the emergency procedure can only relate to digital information that is objectively false, clearly misleading, and manifestly threatens the honesty of upcoming elections or referenda. Further, the diffusion of that information must be artificial or computerized, deliberate, and massive.

Fourth, the new law against information manipulation provides for strengthening media and information literacy. It also asks online platforms to support training in the use of the Internet. Users better trained in digital environment will be more able to detect disinformation and will thus be less influenced by it. Citizens equipped with digital critical thinking will be protected against disinformation spread online, especially that aiming to affect electoral processes.

In conclusion, the French legislation on the fight against the manipulation of information disseminated online before general elections or referenda shows the limits of regulating the dissemination of digital disinformation.

French public authorities, whether the broadcasting regulatory authority or a judge, can enjoy only limited powers against disinformation because they must act within the constraints of the right to freedom of expression. Thus, the Superior Audiovisual Council has not yet suspended, terminated, or refused authorisation to broadcast in conformity with the law of 22 December 2018 whereas the new emergency procedure directed against false information was resorted to only once.

Further, the French legislature should ensure that online platforms respect freedom of expression and can thus not ask them to proactively take down disinformation. For this author, the fight against manipulation of information should not focus on reacting to digital disinformation but on preventing the impact of disinformation on Internet users. Indeed, digital disinformation will continue to exist in democracies respectful of the right to freedom of expression where public authorities are entitled to decide the suspension or suppression of that disinformation in only exceptional circumstances. What should be tackled are the consequences of digital disinformation on the public.

As demonstrated by this paper, one efficient means in addressing the impact of digital disinformation is to increase the transparency of the origin of content spread online. Providing users with details about the motivations and funding sources of digital information as well as about the process behind its dissemination help them to better assess the veracity of that information. As showed by the French experience, in the fight against manipulation of information during electoral campaigns, legislatures could impose on online platforms an obligation to clearly label political advertising. To be effective, information transparency obligations of online platforms should be closely monitored by an independent public authority.

Another efficient means in addressing the effect of digital disinformation is the increase of media and information literacy. If citizens are better trained in the use of digital technology, they will be more able to detect false information online, including that aiming to influence their voting behaviour. Media literacy should be given to all levels of society and life-long because of the speed of technological change. Although online platforms have started to support media and information literacy programmes, those should be preferably provided by independent institutions.

In parallel, States should sustain the establishment of networks of independent fact-checkers to expose digital disinformation. It is necessary to develop tools – particularly the transparency of digital information and media and information literacy – for fostering a positive engagement with fast-evolving technologies and empowering Internet users to discoverer disinformation.

Facilitating the detection by those users of disinformation spread online is the way forward to diminish its impact.

References

Association for Progressive Communications. (2018, March 13). *Content Regulation in the Digital Age Submission to the United Nations Special Rapporteur on the Right to Freedom of Opinion and Expression.*

Bancaud, D. (2019, February 21). Pourquoi l'éducation aux médias est-elle toujours à la traîne en France? *20 Minutes.*

Brattberg, E. and Tim M. (2018, May 23). *Russian Election Interference: Europe's Counter to Fake News and Cyber Attacks.* Carnegie Endowment for International Peace.

Butcher, P. (2021). COVID-19 as a turning point in the fight against disinformation. *Nature Electronics.* 4, pp. 7–9.

Communication from the Commission to the European Parliament, the Council, the European Economic and Social Committee and the Committee of the Regions. (2020, September 30). Digital Education Action Plan 2021–2027. COM(2020) 624 final.

Constitutional Council, Decision no. 71–44 DC, 16 July 1971.

Constitutional Council, Decision no. 2018–773 DC, 20 December 2018.

Court of Cassation, Correctional Chamber. Ruling no. 14–82587, 20 October 2015.

Declaration of the Rights of Man and the Citizen, 26 August 1789.

Decree no. 2019–297 of 10 April 2019, Relating to Information Obligations of Online Platform Operators Promoting Information Content Related to a General Interest Debate.

Dias Oliva, T. (2020). Content Moderation Technologies: Applying Human Rights Standards to Protect Freedom of Expression. *Human Rights Law Review,* 20(4), pp. 607–640.

Draft Articles on Responsibility of States for Internationally Wrongful Acts, with Commentaries. (2001). UN Doc. Res Ass. A/56/83.

Dreyer, E. (2019, January). Fausse bonne nouvelle: la loi du 22 décembre 2018 relative à la manipulation de l'information est parue. *Légipresse,* 367, pp. 19–33.

European Commission High-Level Group of Experts. (2018, March 12). *A Multi-Dimensional Approach to Disinformation.* Report of the Independent High-Level Group on Fake News and Online Disinformation. European Commission.

European Convention on Human Rights, 4 November 1950.

Frassa, C. A. (2018). Opinion No. 667 Made on Behalf of the Commission for Constitutional Laws, Legislation, Universal Suffrage, Regulation and General Administration on the Private Member's Bill Adopted by the Assembly General

After Initiating the Accelerated Procedure, on the Fight Against the Manipulation of Information.

Guillaume, M. (2019, February). Combating the Manipulation of Information – a French Case. *Strategic Analysis*, pp. 1–8.

Jambot, S. (2017, May 29). Macron flingue Russia Today et Sputnik qu'il qualifie d'"organes de propaganda mensongère". *France 24*.

Judgment of the Paris "Tribunal of big instance." 17 May 2019.

La rédaction. (2019, June 4). Fausses nouvelles, manipulation de l'information: comment lutter contre les "fake news"? *Vie publique*.

Land, M. K. (2019). Regulating Private Harms Online: Content Regulation Under Human Rights Law. In R. F. Jørgensen (Ed.). *Human Rights in the Age of Platforms* (pp. 285–316). MIT Press.

Law no. 2004–575 of 21 June 2004 for Confidence in the Digital Economy.

Law no. 2013–1028 of 15 November 2013 on the Independence of the Public Audiovisual.

Law no. 2018–1202 of 22 December 2018 on the Fight Against the Manipulation of Information.

Macron, E. (2018, January 3). New Year's Greeting by President of the Republic Emmanuel Macron to the Press. Elysée.

Mohan, M. (2017, May 9). Macron Leaks: The Anatomy of a Hack. BBC News.

Mounier, J. L. (2019, June 19). Inefficace ou mal comprise, la loi contre les 'fake news' toujours en question. *France 24*.

Moutchou, N. (2018, May 30). General Discussion. Report Made on Behalf of the Commission for Cultural Affairs and Education on the Private Member's Bill. National Assembly. No. 990.

Organic law no. 2018–2101 of 22 December 2018 on the Fight Against the Manipulation of Information.

Report of the Special Rapporteur on the Promotion and Protection of the Right to Freedom of Opinion and Expression. David Kaye. (2016, May 11). UN Doc. A/HRC/32/38.

Report of the Special Representative of the Secretary-General on the Issue of Human Rights and Transnational Corporations and Other Business Enterprises, John Ruggie. (2011, March 21). Guiding Principles on Business and Human Rights: Implementing the United Nations 'Protect, Respect and Remedy' Framework. UN Doc A/HRC/17/31 (Annex).

Superior Audiovisual Council. (2020, July 30). *Lutte contre la diffusion de fausses informations sur les plateformes en ligne*.

Superior Audiovisual Council. (2019). Recommendation no. 2019–13 of 15 May 2019 of the Superior Audiovisual Council to Online Platform Operators in the Context of the Duty to Cooperate to Fight the Dissemination of False Information.

Studer, B. (2018, May 30). General Discussion. Report Made on Behalf of the Commission for Cultural Affairs and Education on the Private Member's Bill. National Assembly. No. 990.

Vilmer, J. B. et al. (2018). *Information Manipulation: A Challenge for Our Democracies.* Report by the Policy Planning Staff of the Ministry for Europe and Foreign Affairs and the Institute for Strategic Research of the Ministry for the Armed Forces.

CHAPTER 7

Taming Our Digital Overlords: Tackling Tech through 'Self-Regulation'?

Kim Barker

In the era of user-generated content, our ability to determine what is lawful, trustworthy, and legitimate has never been more important. Increasing usage of, reliance on, and engagement with social media and user-generated content platforms has created something of a 'content crisis' by which the content consumed has exponentially increased in volume while not per se developing in accuracy or reliability. This 'infodemic' (World Health Organisation, 2020) poses difficult questions for regulatory and governance mechanisms, but content is now a cornerstone of our digital society.

While the user-generated content economy grows at an uncontrollable rate, regulatory bodies, governments, and policymakers continue to grapple with questions of how to govern the platforms hosting (and encouraging) this content. Not only does harmful and / or illegal content prove problematic for those seeing (and relying) on it but categorising it for action has remained an elusive ambition. The 'infodemic' and the illegal content epidemic combine to present a unique juxtaposition, but also a particularly individual set of governance dilemmas. Information has never before been so readily accessible, and at the same time, it has rarely been so challenging to control.

The argument explored here positions founders and operators as the custodians of their platforms, but also overlords of the internet. In placing these at the heart of the discussion, the acquiescence of lawmakers is critiqued, as is the development and reliance upon self-regulatory mechanisms for dealing with online platforms and their responsibilities in the regulation of online content. This paper advances the critique that too much attention has been paid to categorising content to the detriment of a broader regulatory regime. The argument explored here focuses on the underappreciated but central roles of self-regulation actively pursued as the cornerstone of platform regulation before assessing the scope for improving the accountability of self-regulatory mechanisms through the concept of trust as something online platforms themselves have worked to develop. The discussion here ultimately concludes that the trend to self-regulatory mechanisms remains, but there ought to be a shift in focus to regulate oversight groups, rather than platforms. The discussion

 | DOI:10.1163/9789004547261_009

therefore turns first to an assessment of the platform ‘problem’, and the role of digital overlords in (failing) to tackle the content crisis.

1 The Platform ‘Problem’ – Digital Overlords

Public perceptions, content media, and online information are all prevalent, everyday norms of engaging in a digital world. The ways in which we interact – and rely – on platforms, but especially social media platforms, present significant challenges to contemporary life and participation. These challenges arise not just in respect of the well-recognised concerns over what is shared and by whom via such platforms, but also in respect of the volume of posts and the need for rapid responses to problematic content. Further concerns arise in respect of the types of information shared through social media platforms, particularly where that information is misleading, harmful, or unlawful. There are other connected issues that feed in to – what is called here – ‘the platform problem’, and they relate specifically to the trust that users are asked to place in platform operators to control their spaces by lawmakers and the legal institutions.

Platforms – such as e.g., Twitter, Facebook, Instagram, and TikTok – have developed, partially by accident but partially by design, to operate in a manner which makes legal oversight and regulation less than straightforward. In developing as start-up entities, often with little thought or planning given to their legal and operational status (Levy, 2020) and with a focus falling on the programming or technical operations of their online sites (Losse, 2012), not much attention has been paid to the rules concerning how users of each of these sites should be allowed to behave when using the site. Terms, conditions, usage agreements, and policies have been developed as an afterthought, in much the same way as legal regulation has developed in a knee-jerk manner. In part, the speed of growth and evolution of these platforms has contributed to this, and the law has struggled to maintain an understanding of the legal issues, let alone of the ever-evolving regulatory needs. This is compounded by the evolution of different types of platforms, which have – in turn – spurned different types of content and user activities. The rapidity of such technical developments poses future-proofing obstacles for law-making – something that has very much come to the fore in recent years with the explosion in online content generation, but also the need to moderate / review such content.

By virtue of the piecemeal evolution of social media platforms, the operators of such entities have found themselves in positions whereby they have a direct level of control over the speech and expression abilities of individual users. The platform founders and controllers have become – to borrow Levy's phrase – the "arbiters of speech" (2020: 444) online, or "custodians of the Internet" (Gillespie, 2021). The group of individual start-up founders have – inadvertently – become the de facto rule-setters, and the enforcers of rules across their own platforms. Their influence spreads more widely than that – these individuals have also become the establishers of Internet norms for the posting and sharing of content but also for setting the parameters of what should be shared without recourse and things which can be shared but which have legal consequences.

Lawmakers and legal systems have been complicit in elevating the platform providers to positions of digital 'overlords', empowering them through legal acquiescence to become the decision makers in online spaces, or legislator, prosecutor, jury, and judge all in one. By failing to appreciate the speed and spread of problematic content and actions online, those charged with assessing and reporting on a need to regulate the internet have allowed online platforms to accrue power and influence almost unabated. Content problems are no longer containable and cannot be addressed through solely delegating responsibility to private business entities with vested interests in both the spaces *and* the control of them.

For instance, of the dominant platforms, few – if any – have not faced regulatory difficulties or criticism for failing to tackle issues that have arisen on their sites. For Twitter, it has been the outpour of vitriolic, abusive, and hateful tweets (Amnesty International, 2018). Facebook too has encountered numerous challenges, from leaving explicit and terror-related content on its site (ABC, 2019), to over-zealously (and wrongly) removing content designed to raise awareness of breast cancer (BBC News, 2016) and breastfeeding (Sweney, 2008). These platforms are not alone in facing content challenges though – TikTok too has faced very recent but challenging content crises, notably the graphic beheading video which duped viewers into believing they were watching a dance video before the clip cut to the extremely explicit content (Fowler, 2021). These are just selected examples of the difficulties platforms face not only in regulating the content which is posted on their sites but also in determining how to deal with that content when it becomes problematic or violates either their own, respective rules and standards, or the law (Barker and Jurasz, 2021).

Google and Amazon – categorized here as consumer platforms – have faced different criticisms, relating to the abuse of their dominance in their respective markets. Google for instance has been fined three times for abusive practices, with the most recent coming in 2019 for abusive advertising practices it used while operating as an intermediary (European Commission, 2019). Amazon, meanwhile, is under investigation in Canada and the European Union for potentially anti-competitive practices on its e-commerce site (Bolongaro, 2020). While these latter concerns do not relate specifically to the issues of content and fall into the category of platform behaviours rather than user behaviours, they do nonetheless serve as reminders of the ways in which online platforms (social and commercial) operate, and the ways in which they have grown to dwarf regulatory mechanisms and, arguably, competence.

The peculiarities as well as shared similarities across all online operators pose additional challenges to potential regulatory mechanisms. All of the platforms engage with content – from users and from their own in-house content teams – but the intended uses of this content is very different. That distinction has, it is suggested here – coloured the ways in which governance discussions have unfolded and has led, recently – at least in Europe – to a differentiated regime being proposed to address contemporary challenges posed by online platforms.

2 Tackling Technology: a Self-Regulatory Agenda

In struggling to address online platform behaviours, governance, regulation, and legal mechanisms have all encountered significant obstacles. Part of the overarching difficulty arises with the conceptualization and resulting categorization of these platforms. They all share some features – for instance, they are all online, and rely on the internet to operate and attract a user base. They are all also very heavily focussed on user behaviours, from sharing and posting, to liking content, to engaging with content and purchasing based on it. There are, however, some distinctions which also present challenges when attempting to introduce governance mechanisms. For instance, social media platforms are reliant upon users posting, sharing, recommending and liking content to drive other users to the site and to specific forms of content. The algorithmic feeds of these sites are engineered around this. Other platforms – such as commercial platforms operate in similar ways, relying upon things which have been viewed, or liked, or which are popular to determine what may also be of interest to users for purchase. Irrespective of their mode of operation, online platforms have generated significant discussion, and questions of governance continue to be plentiful.

There has been a wealth of discussion in Australia, the UK, Europe – including Germany, France, Austria and Hungary – as well across the United States of America in recent years concerning the role and responsibilities placed on online platforms, but especially social media platforms. The agenda for such discussions has been set largely by the rise in their prolific use, combined with the wealth of content generated and posted every 24 hours, but more especially, the impact such accessible and replicable and potentially harmful content can have. Discussions across Europe have been similarly triggered by the passage of time – it is now over 20 years since the European Commission introduced the eCommerce Directive, as one of the first initiatives to grapple with the protections and liability for hosting platforms. This anniversary, together with the rise of 'problematic content' including hate speech, intellectual property infringement, disinformation and misinformation, online abuse, online violence against women, and extremist content, have combined to offer the potential for a shift in the regulatory landscape. This opportunity has resulted in significant political will, and legislative endeavour at regional, as well as at domestic levels. As the European Commission Vice President states, "Now we have such an increase in the online traffic that we need to make rules that put order into chaos" (Scott, Thibault and Kayali, 2020). The 'ordering' that Vestager refers to here is the Digital Services Act (DSA) 2020, which is the much-vaunted replacement to the eCommerce Directive, but one which goes beyond the original principles. The DSA is intended to "define clear responsibilities and accountability for providers of intermediary services, and in particular online platforms, such as social media" (European Commission, 2020: 2). The statement by Vestager is a prescient observation, but it is also a damning assessment of the current rules, and their current obsession with categorising content into lawful and unlawful – something of a historic and pre-user generated content – legacy.

3 Tentatively Talking Self-Regulation

In responding to the need to establish 'order' (Scott, Thibault and Kayali, 2020), there have been – historically – a plethora of rules and approaches considered to tackle online content. The initial efforts that were introduced to regulate online platforms and services conducting their business online adopted rules that placed two things at the core of the legal regime: (i) actual knowledge of illegal activity, and (ii) expeditious action in the event of actual knowledge. These concepts were established in Article 14 of the eCommerce Directive (2000).

The same legislative instrument also introduced notions of what Kuczerawy calls "self-regulation" (2017: 227) and which are found within Article 16 of the eCommerce Directive itself. The legislative discretion for enacting states allowed for the *possibility* of legislating to allow platforms to introduce self-regulation measures within national jurisdictions, but very few such systems were actually introduced despite such scope being enshrined in the eCommerce Directive. While there was – in the late 1990s – some merit to thinking that states would introduce self-regulatory measures to tackle online platforms and online content, this fell flat, with the overwhelming majority of enacting states simply transposing the eCommerce Directive provisions verbatim (Kuczerawy, 2017: 227; van Eecke and Truyens, 2009: 19). This was of course coupled with the prohibition on any general monitoring obligation, meaning platforms were required to do very little proactively.

The eCommerce Directive is therefore intriguing for two connected reasons. Firstly, it was the original legislation that required action to tackle online content – albeit not proactively. Secondly, it expressly encouraged – by virtue of Article 16 – a self-regulatory approach. In essence, this has put platform operators and the platforms themselves in significant positions of responsibility. Provisions for legislative action, meanwhile, go no further than allowing for national states to introduce self-regulatory mechanisms, essentially placing the responsibility and the burden at the door of those who conceived of the platforms to begin with – but who had, notably, not conceived of ways to control their own spaces.

In sum therefore, the first Europe-wide initiative aimed at addressing online regulation of content was focussed on protecting platforms hosting content and did not require any monitoring while at the same time – optimistically – allowing for states to go further and introduce aspects of self-regulation. This combination offers what is in essence a legal shield for platforms from any responsibility addressing online content. This is – in part – perhaps due to the concerns surrounding freedom of expression rights, and the dangers of over-regulating speech in online contexts. That said, the eCommerce Directive has laid the foundations for a legacy of protection being given to platforms. In turn, this has led – arguably – to situations where platforms are neither encouraged nor incentivised to do more than the basics when it comes to controlling, reviewing, and tackling problematic content on their sites. This attitude has, for the most part, been the prevailing one in the period between the introduction of the eCommerce Directive, and the proposed Digital Services Act (DSA) 2020.

4 The Renewal of Self-Regulation

In the absence of targeted legislative models for online material, reliance falls on general legal rules which are not tailored specifically for the internet – one of the most significant issues identified by the Council of Europe after the eCommerce Directive was introduced, but before its replacement had been proposed (CoE, 2017: 3). In introducing regulatory standards – albeit minimum standards – the eCommerce Directive established that there was some need to address the gap across Europe. Its introduction, therefore, signalled something of a shift and highlighted the emergence of a new era of regulation beyond individual platforms dabbling with rules on their specific sites. It also sought to put platforms into positions where there was quasi-official recognition of their role in regulating online speech. In some respects, the eCommerce Directive introduced an additional and much-needed layer of responsibility beyond a contractual set of terms and conditions or non-platform specific means of regulation (Barker, 2016: 64). It did not, however, introduce, nor offer, what de Streel and Huvosec refer to as a "comprehensive tool of removal of illegal content" (2020: 10). While fair criticism of the eCommerce Directive, it was conceptualised at the beginning of the online speech era, when few rules and legal provisions existed to capture online content (York, 2021:11), let alone regulate its acceptability (or otherwise). Given these limitations, it is unsurprising that the eCommerce Directive is a blunt tool.

Various assessments of the Directive have highlighted – as social media and other online platforms grow in importance – the need for additional guidance. In 2016–2017, recognition came from the European Union that much more encouragement was required for self-regulating illegal materials online that are regarded as harmful (de Streel and Huvosec, 2020: 9). This again reiterates the importance of self-regulatory tools in the current regime. This reminder also serves to highlight how little attention was paid to such governance mechanisms by individual platforms, reiterating the difficulties in this area of law. Historically, emphasis was placed on how to bring in the platform operators themselves to play a part in governing online content, rather than seeking to rely on national or regional oversight measures.

Post-2016, some tentative steps towards an enhanced level of platform responsibility have been made, albeit at a European level. For instance, the Code of Conduct on Illegal Content Online was introduced in 2016 and had the support of – at the time – four major online platforms including Facebook and Twitter (European Commission, 2016). While a voluntary Communication from the European Commission, it signalled a new era of greater involvement by – and with – online platforms. In some respects, therefore it can be heralded as an initiative which has taken proactive steps to engage platforms in self-regulatory mechanisms. Whether this is quite what was envisaged in the eCommerce Directive is unlikely, not least because the Directive anticipated self-regulatory mechanisms at a state level, rather than a regional one.

The transposition flexibility left for member states to address in respect of self-regulation suggests that the Code of Conduct approach at a regional level was something different, but nonetheless, a start on self-regulation. The significance of this should not be overlooked – the involvement of the platforms, even if a small number initially, also highlights that there has been a shift in their willingness to engage, but also – perhaps – an acknowledgement that they have a role to play in the governance of their sites, and of online content more broadly.

This is certainly something which was witnessed with Zuckerberg and Facebook in 2019 when he claimed to want more regulation for technology (Zuckerberg, 2019) – something which caught politicians by "surprise" (Stolton, 2019). In calling for "new regulation", Zuckerberg laid out his vision for what he referred to as "a common global framework" instead of piecemeal elements of governance (2019). While an interesting signal, his calls came at a time when internet regulation and law reform addressing the internet was already on the horizon in Europe, the UK, Australia, Canada, and the USA to name but a few states. Zuckerberg's call was also something of a stark contradiction to some ideas suggested at domestic levels in European countries – with German Chancellor Angela Merkel suggesting there was a need for Europe to have its own internet, with its own rules, what she referred to as a "European communications network" (Miller, 2014) – something reminiscent of China's stance regarding the internet, and the Chinese Great Firewall (Economy, 2018).

Irrespective of the differences of opinion between platform founders, and European leaders, Zuckerberg's call was surprising, not least because of his previous reticence to be held to account (Griffin, 2018; Stolton, 2019). It did, though, lead to a fuller discussion of Facebook's ideas for how content regulation could develop, with Facebook's White Paper having been published in 2020 (Bickert, 2020) – a document considering principles for regulators. In producing such a document, it seems that Facebook is attempting to shape and define the obligations that may be imposed on it by lawmakers anyway – something which is a far cry from the origins of a platform which developed without much attention being paid to what users could and could not do. It is, perhaps, also a further indicator that social media platforms are now prepared to play a more involved role in addressing online content problems. If not, then it is, at the very least, recognition of the inevitable – lawmakers and regulators are seeking to change the governance framework for online platforms, whether platforms are willing to engage with that ambition or not.

Other European measures have followed the Code of Conduct, including a series of workshops aimed at addressing fundamental rights, notice and takedown mechanisms, and voluntary measures (European Commission: 2017a, European Commission: 2017b, European Commission, 2017c). These have been supplemented with regular reporting on the effectiveness of self-regulatory measures, suggesting that there has been significant progress, something that the Council of Europe is very keen to highlight. In its most recent assessment, it comments that the Code of Conduct has "contributed to achieve quick progress … It has increased trust and cooperation" (CoE, 2019: 2). There does, however, remain a significant element of concern over the roles of private entities in overseeing speech and expression, irrespective of the claims surrounding progress made in trust building. And above all else, these self-regulatory measures have little legal basis – they are the outcome of what is described, at best, as cooperation with the European Commission, but little else. That said, they are indicators of a willingness by platforms to get involved, and address some of the issues posed by the content they encourage and host. This shift towards taking some responsibility is encouraging because it suggests that there is acceptance by the platforms that the eCommerce Directive approach is no longer one fit for the arena in which it is asked to operate. This would also – at least in part – explain the shift in approach from Facebook. A generous assessment would also indicate that Facebook, like other platforms, has evolved somewhat, and recognised that the problems posed by speech and content cannot go unchecked, nor can engaging with governance mechanisms be ignored.

Despite the progress, there remain serious flaws with the Code of Conduct, initiatives like it, and the eCommerce Directive. Fundamentally, as well-intended as self-regulation for online content is – and it was, ostensibly, based on other media industry regulation models – it rests on the notion of voluntariness. The nature of the Code of Conduct was voluntary in name only – in contrast to other codes of conduct at a regional level, the European Commission was the driving force behind this, and not only that, but also involved a monitoring system to ensure that there was implementation. This is "markedly different" from a truly voluntary system (Bukovská, 2019: 3), suggesting that even in 2016, the European Commission had given up on the notions of true self-regulation envisaged at the turn of the millennium. This would seem to be the trend, particularly with the signals sent by Vestager of needing to introduce rules to tackle the online environment, and with the Digital Services Act proposal coming forward in late 2020 – a direct replacement for the eCommerce Directive. These changes, while reflective of a shift in emphasis at a European level, also reflect a greater trend towards statutory regulation, and away from self-regulatory mechanisms.

5 Towards Statutory Intervention

As a tentative step away from self-regulation, the follow-up measures to the Code of Conduct indicated a greater appetite at a European level to engage proactively in the regulation of online content, and of online platforms. These tentative measures, despite being hailed as progressive (CoE 2019: 2), were – it transpires – just the beginning of what is a much bolder ambition to introduce more appropriate governance systems for online platforms. This has become evident in individual European states since 2017, and mirrors broadly the greater interest taken at a regional level around the same time.

In Germany for example, discussions about national internet regulation and online hate speech laws quickly turned to criticisms of the Netzwerkdurchsetzungsgesetz 2017 (NetzDG) (Network Enforcement Act) which was introduced to make online platforms liable for illegal content which is not deleted within a short period of time. The NetzDG is not without significant controversy, but has been heralded as successful, with at least 13 other states using it as a model for similar provisions at a national level (Mchangama and Fiss, 2019). Not only has it gained significant plaudits, as well as outcry, but it has signaled that there is a sea change in the ways in which online platforms, especially social networks are going to be regulated going forward. While the NetzDG is designed to tackle illegal content online according to German law, it has prompted significant criticisms over the likely over-removal of speech (Tworek and Leerssen, 2019: 3), yet tellingly, does not shift the emphasis away from platforms. Instead, despite suggesting that it will change the dynamic of online regulation and impose a regime that does not place platforms at the heart of decision making, platforms are once again charged with privatized enforcement because they are placed in positions where they are required to determine the legality of content without judicial oversight (HRW, 2018). In essence, despite its signals, and the controversy, the NetzDG does little more than reinforce the role of platforms in self-regulatory mechanisms, but with greater potential punishments for falling short of the statutory expectations – a recurrent theme in national laws.

Germany is not the only state where national legislation has been tabled to tackle online platforms. It has been used a model in Austria through the Communication Platforms Act 2020 (KoPL-G), in France through Bill No 1785 Aiming to Fight Against Hatred on the Internet ('Avia' Bill) 2020, in Poland through the Freedom of Speech Act, and latterly in the UK through the proposed Online Safety Bill 2021. The majority of the domestic provisions seek to do little more than make online platforms responsible for not acting to address illegal content. The UK Bill, for instance, proposes something similar but with a different focus, and instead proposes to impose a 'duty of care' on online platforms under section 5 of the Bill. This 'duty' will apply in respect of an indicative list of 23 types of online harms identified during the preparatory and consultative phases (HM Government, 2020). The list is not exhaustive, and omits contemporary harms including online violence against women, to name but one (Barker and Jurasz, 2021b). Nevertheless, it too shies away from taking steps that move beyond elements of self-regulation, albeit does propose that there is an oversight body empowered by statute. In some respects, therefore, this domestic attempt at governing online platforms too is constrained by what has gone before it and retains notions of platform responsibility seen elsewhere.

All of these legislative developments have progressed against the regional backdrop of the European Digital Services Act 2020 (DSA) proposal, which also aims to change the way in which online platforms are regulated. The DSA does little different. In fact, it seeks to replicate – and retain – the provisions of the eCommerce Directive. In so doing, it keeps, but arguably strengthens, the liability shield for online platforms, and seems to have responded to platform providers' requests to retain the eCommerce Directive regime unchanged (EU Commission, 2017c: 3). It is therefore not as radical as it could have been. This in itself is reflective of the tension between the regional and national appetite for regulation. The DSA – perhaps – has been drafted in the full knowledge that the self-regulatory scope embedded in the eCommerce Directive was not acted upon at a national level. Given that this was the situation, it is of little surprise that the DSA does not seek to make wholescale radical changes. Yet at the same time, it is surprising that changes have not been made to the liability shield, nor to elements of self-regulation given the plethora of national initiatives indicating that national legislative measures are being pursued to address the governance of online platforms. It is surprising that the regional initiative does not do more to tackle the desired responsibility of online platforms in manners that national initiatives – at least objectively – wish for.

6 Better, Not More: Taming the Overlords

In continuing to pursue self-regulation as the predominant mode of governance, little appears to be substantively changing, despite the numerous legislative initiatives. What is evident, is that aside from seeking to alter governance approaches, there is a drive to introduce more comprehensive sets of obligations and provisions designed to improve some of the lack of transparency that has crept into statutory approaches to online platforms.

As a result, while there is a drive for more regulation, there is essentially a campaign for better regulation too. That does not actually address the central problem though, and introducing more legislation, especially disparate legislation across numerous jurisdictions, is unlikely to assist online platforms in their self-regulatory obligations. It is yet to be determined whether – if at all – any of these new legislative initiatives improve the situation in respect of governing online platforms. It is to be hoped that the numerous provisions do not undo the ambition of improving the regulatory framework or tackling the digital overlords.

Aside from the volume of legislative proposals currently on the horizon both regionally, and nationally, there remains the ultimate question of how to better address the overarching issue of governing online platforms. What each of the national proposals seeks to do, is enhance the punishments available for not acting to address illegal content online, without necessarily improving the transparency and / or guidance around how platforms are expected to tackle that. Some of the proposed legislation seeks to introduce a different arrangement, such as the UK's Online Safety Bill for instance. Yet that too, still seeks to impose hefty financial punishment for not acting, or not acting within the required timescales. In that respect, few of the proposals are substantively different in their overall focus from the eCommerce Directive, and none depart from the notion of self-regulation, despite purporting to be making online platforms more accountable.

For instance, few of the proposals in the various legislative initiatives talk about the responsibilities placed on online platforms, and even fewer mention duties. There is – it is suggested here – a missing aspect to all of these discussions given the emphasis that continues to fall on self-regulation – trust. All of the proposals and self-regulatory mechanisms implicitly rely upon being able to trust online platforms to act appropriately within the parameters set – either legislatively or through ostensibly voluntary Codes of Conduct. As the gatekeepers to the internet, and our digital overlords charged with oversight of online speech, platform providers are being placed in positions where they are determining what can and cannot be said. Ostensibly legal standards are being upheld, but certainly at a European level there are a number of flaws in the approach adopted, several of which relate to the lack of a normative legal basis for addressing various facets of speech (Bukovská, 2019: 3). Ultimately, this is concerning, not least because by positioning the platforms as decision-makers, users and legislators are placing trust in them to act in accordance with the anticipated legal standards, often with little judicial oversight. In continuing with self-governance mechanisms and allowing platforms and media outlets to publish content without responsibility or verification, is to continue to rely on systems that endanger freedom of expression and rely on legal systems designed to filter content, or to takedown content.

Online platforms have themselves started to evolve measures designed to capture and build notions of 'trust'. For instance, Twitter has established its own Trust and Safety Council (Cartes, 2016), deliberately implanting ideas of trustworthiness in the oversight arrangements it has introduced for its own platform. It is not alone, Facebook too has acted to introduce elements of control and review that are designed to remind users that it is an entity in which one can place trust – but unlike Twitter, Facebook has not opted to use 'trust' as a phrase, instead it is relying on pre-existing trustable entities such as courts to engender notions of legitimacy. In establishing its Oversight Board (Clegg, 2020), to act as a 'quasi-court', it is suggesting that it is trustable, despite the obvious concerns over self-interest given the ways in which it is funded by Facebook itself. Both of these initiatives are designed to message – both overtly and subliminally – trust within the platforms, but also across the user base. They are also neat messages for legislators as to the proactivity with which governance mechanisms are being pursued – perhaps to suggest that further legal regulation is not necessary, despite Zuckerberg's overtures to the contrary in 2019 and 2020. In some respects, therefore, with the European Commission, online platforms, and civil society organisations all invoking notions of trust, it is time to consider the role of trust within the self-regulatory and governance landscape. Online platforms seem to have shown willingness to hold themselves to account through mechanisms of their own design. If these are to be trusted and relied upon, governance mechanisms should recalibrate to examine how these mechanisms could be governed instead of focussing on the regulation of online platforms as a one-size fits all effort.

In establishing platform-specific oversight bodies, an additional layer of self-regulation has been introduced, albeit one which was not necessarily anticipated nor envisaged by legislators. This in itself makes governance of these oversight bodies worth considering as one way in which our digital overlords could be held to account. Trust, as something that platforms have initiated themselves as a concept and a tool, should be placed front and centre of governance discussions given the trend towards self-regulation which seems set to continue in light of the DSA's reaffirmation of that approach. In some respects, the potential for self-regulation to be an appropriate and reliable mechanism takes on greater significance with the advent of individual oversight bodies suited to the workings of each unique platform – it is a demonstration of a platform seeking to police itself separately from the obligations imposed on it by legal mechanisms. That should mean that there is a need to take it seriously, and, if self-regulation is the preferred governance approach from legislators, to refocus efforts on regulating the oversight groups, rather than the platforms themselves. It would also allow for the development of tailored regulation per platform, rather than the nefarious and granular difficulties of categorising content into lawful, lawful but harmful, and unlawful. Instead of diluting governance discussions with obscure notions of duties of care, trust is a concept that is recognisable legally, and non-legally, but one which could allow for the taming of digital overlords when it comes to online regulation. It would also fill the void between legislative notions of self-regulation and platform initiated self-accountability fora.

7 Conclusion: Recalibrating Self-Regulation?

The problems posed by platforms are multifaceted but compounded by the ways in which platform operators have been placed at the heart of the legal framework that has been – historically – envisaged to address these issues. That strategy has, arguably, run its course, leaving a responsibility vacuum within the regulatory sphere. The platform operators, as our digital overlords, are both part of the problem and potentially part of the solution. Legal regulation, or the outsourcing of it, to platform operators can no longer be an appropriate one-size-fits-all solution. It is time to recalibrate our thinking of online platforms, but also their regulation.

Putting trust at the heart of this recalibration can offer different, potentially more effective solutions to the problems social media platforms and digital technologies will continue to pose. There is a clear desire for governance of platforms, and platforms themselves indicate they are at least willing to engage in this process. By shifting the narrative and adopting a holistic trust-centric approach, the thorny issues surrounding online content and platform regulation can be addressed, and effect be given to the long-held European ambition of self-regulation for platforms. By positioning our digital overlords as trustworthy, the emphasis, like the governance mechanisms built around such notions, could become workable.

References

ABC. (2019, 19 March). New Zealand PM Jacinda Ardern leans on Facebook to drop Christchurch shooting footage. *ABC News*. Retrieved from: https://www.abc.net.au/news/2019-03-19/new-zealand-facebook-christchurch-shooting-video-sheryl-sandberg/10915184.

Amnesty International. (2018, March). Toxic Twitter. Retrieved from: https://www.amnesty.org/en/latest/research/2018/03/online-violence-against-women-chapter-1/.

Barker, K. (2016). 'Virtual spaces and virtual layers – governing the ungovernable?' *Information & Communications Technology Law*. Vol 25(1), 62–70. DOI: https://doi.org/10.1080/13600834.2015.1134146.

Barker, K. and Jurasz, O. Online Misogyny as a Hate Crime: An Obstacle to Equality? *GenIUS* (forthcoming, 2021a).

Barker, K. and Jurasz, O. (2021b). Text-Based (Sexual) Abuse and Online Violence Against Women: Toward Law Reform? In: Bailey, J., Flynn, A., and Henry, N., eds., *The Emerald International Handbook of Technology Facilitated Violence and Abuse*. [online] Bingley: Emerald Publishing Limited, pp. 247–264. Available at: https://doi.org/10.1108/978-1-83982-848-520211017.

BBC News. (2016, 20 October). Facebook apologises for removing breast cancer awareness video. Retrieved from: https://www.bbc.co.uk/news/world-europe-37721193.

Bickert, M. (2020, February). Charting a Way Forward: Online Content Regulation. *Facebook*. Retrieved from: https://about.fb.com/wp-content/uploads/2020/02/Charting-A-Way-Forward_Online-Content-Regulation-White-Paper-1.pdf.

Bukovská, B. (2019, 7 May). The European Commission's Code of Conduct for Countering Illegal Hate Speech Online. An analysis of freedom of expression implications. ARTICLE 19. Retrieved from: https://www.ivir.nl/publicaties/download/Bukovska.pdf.

Bill No 1785 Aiming to Fight Against Hatred on the Internet ('Avia' Bill) 2020.

Bolangaro, K. (2020, 14 August). Amazon 'Abuse of Dominance' Concerns Trigger Probe in Canada. *Bloomberg*. Retrieved from: https://www.bloomberg.com/news/articles/2020-08-14/canada-competition-bureau-probes-amazon-for-abuse-of-dominance-kdudbgwl.

Cartes, P. (2016, 9 February). Announcing the Twitter Trust & Safety Council. Retrieved from: https://blog.twitter.com/en_us/a/2016/announcing-the-twitter-trust-safety-council.

Clegg, N. (2020, 6 May). Welcoming the Oversight Board. Retrieved from: https://about.fb.com/news/2020/05/welcoming-the-oversight-board/.

Communication Platforms Act 2020. (KoPL-G).

Council of Europe (2019, 27 September). Assessment of the Code of Conduct on Hate Speech on line *State of Play*. (12522/19). Retrieved from: https://ec.europa.eu/info/sites/default/files/aid_development_cooperation_fundamental_rights/assessment_of_the_code_of_conduct_on_hate_speech_on_line_-_state_of_play__0.pdf.

Council of Europe. (2017, January). Comparative Study on Blocking, Filtering and Take-Down of Illegal Internet Content. *Council of Europe*. Retrieved from: file:///Users/KimB/Downloads/014017GBR-Illegal%20Internet.pdf.

Directive 2000/31/EC of the European Parliament and of the Council of 8 June 2000 on Certain Legal Aspects of Information Society Services, in Particular Electronic Commerce, in the Internal Market ('Directive on Electronic Commerce') [2000] OJ L 178/1. (eCommerce Directive).

Economy, E. (2018, 29 June). The great firewall of China: Xi Jinping's internet shutdown. *The Guardian*. Retrieved from: https://www.theguardian.com/news/2018/jun/29/the-great-firewall-of-china-xi-jinpings-internet-shutdown.

European Commission. (2019, 20 March). Antitrust: Commission fines Google €1.49 billion for abusive practices in online advertising. Retrieved from: https://ec.europa.eu/commission/presscorner/detail/en/IP_19_1770.

European Commission. (2016, 30 June). Code of conduct on countering illegal hate speech online. Retrieved from: https://ec.europa.eu/info/policies/justice-and-fundamental-rights/combatting-discrimination/racism-and-xenophobia/eu-code-conduct-countering-illegal-hate-speech-online_en.

European Commission. (2017a, 28 September). Communication from the Commission to the European Parliament, The Council, The European Economic and Social Committee and Committee of the Regions on Tackling Illegal Content Online – Towards an Enhanced Responsibility of Online Platforms. COM(2017) 555 final. Retrieved from: https://eur-lex.europa.eu/legal-content/EN/TXT/?uri=CELEX%3A52017DC0555.

European Commission. (2017b). Evidence gathering on liability of online intermediaries. *European Commission*. Retrieved from: https://digital-strategy.ec.europa.eu/en/library/evidence-gathering-liability-online-intermediaries.

European Commission. (2020, 15 December). Regulation of the European Parliament and of the Council on a Single Market for Digital Services (Digital Services Act) and amending Directive 2000/31/EC. Retrieved from: https://eur-lex.europa.eu/legal-content/EN/TXT/PDF/?uri=CELEX:52020PC0825&from=EN (DSA).

European Commission. (2017c, 6 April). Workshop on Voluntary Measures. Retrieved from: https://digital-strategy.ec.europa.eu/en/library/evidence-gathering-liability-online-intermediaries.

van Eecke, P. and Truyens, M. (2009, November). 'Legal Analysis of a Single Market for the Information Society, New Rules for a new age?' A study commissioned by the European Commission's Information Society and Media Directorate-General. *Council of Europe*.

Fowler, K. (2021, 9 June). TikTok Faces Huge Test as Graphic Video Floods Platform Popular With Teens. *Newsweek*. Retrieved from: https://www.newsweek.com/tiktok-graphic-video-beheading-problems-users-ai-moderation-system-1598844.

Gillespie, T. (2021). *Custodians of the Internet. Platforms, Content Moderation and the Hidden Decisions that Shape Social Media*. Yale University Press: New Haven.

Griffin, A. (2018, 31 October). Facebook's Mark Zuckerberg Summoned to Appear Before UK and Canadian Parliaments. *The Independent*. Retrieved from: https://www.independent.co.uk/life-style/gadgets-and-tech/news/mark-zuckerberg-parliament-uk-canada-westminster-dcms-select-committee-a8609996.html.

HM Government, 'Online Harms White Paper: Full government response to the consultation' (2020) Cmd 354.

Human Rights Watch. (2018, 14 February). Germany: Flawed Social Media Law. Retrieved from: https://www.hrw.org/news/2018/02/14/germany-flawed-social-media-law.

Kuczerawy, A. (2017). The Power of Positive Thinking: Intermediary Liability and the Effective Enjoyment of the Right to Freedom of Expression. *Journal of Intellectual Property, Information Technology and E-Commerce Law*. Vol 8(3) pp 226–237. https://www.jipitec.eu/issues/jipitec-8-3-2017/4623.

Levy, S. (2020). *Facebook: The Inside Story*. Penguin Random House: UK.

Losse, K. (2012). *The Boy Kings. A Journey Into the Heart of the Social Network*. Free Press: New York.

Mchangama, J. and Fiss, J. (2019, November). The Digital Berlin Wall: How Germany (Accidentally) Created a Prototype for Global Online Censorship. *Justitia*. Retrieved from: http://justitia-int.org/wp-content/uploads/2019/11/Analyse_The-Digital-Berlin-Wall-How-Germany-Accidentally-Created-a-Prototype-for-Global-Online-Censorship.pdf. .

Miller, R. (2014, 17 February). Germany's Merkel calls for Separate European Internet. *Data Center Knowledge*. Retrieved from: https://www.datacenterknowledge.com/archives/2014/02/17/germanys-merkel-calls-separate-european-internet.

Ministerstwo Sprawiedliwości, 'Zachęcamy do zapoznania się z projektem ustawy o ochronie wolności użytkowników serwisów społecznościowych' (1 February 2021) https://www.gov.pl/web/sprawiedliwosc/zachecamy-do-zapoznania-sie-z-projektem-ustawy-o-ochronie-wolnosci-uzytkownikow-serwisow-spolecznosciowych. (Freedom of Speech Act).

Network Enforcement Act (*Netzwerkdurchsetzungsgesetz*) 2017. (NetzDG).

Online Safety Bill 2021.

Scott, M., Thibault, L., and Kayali, L., (2020, 15 December). Europe rewrites rulebook for digital age, *Politico*. Retrieved from: https://www.politico.eu/article/europe-digital-markets-act-services-act-tech-competition-rules-margrethe-vestager-thierry-breton/.

Stolton, S. (2019, 3 April). 'Regulation will not solve Facebook's problems' Commission says. *Euractiv.com*. Retrieved from: https://www.euractiv.com/section/data-protection/news/regulation-will-not-solve-facebooks-problems-commission-says/.

de Streel, A. and Huvosec, M. (2020, May). The ecommerce Directive as the cornerstone of the Internal Market: Assessment and options for reform. *European Parliament*. Retrieved from: https://www.europarl.europa.eu/RegData/etudes/STUD/2020/648797/IPOL_STU%282020%29648797_EN.pdf.

Sweney, T. (2008, 30 December). Mums furious as Facebook removes breastfeeding photos. *The Guardian*. Retrieved from: https://www.theguardian.com/media/2008/dec/30/facebook-breastfeeding-ban.

Tworek, H. and Leerssen, P. (2019, 15 April). An Analysis of Germany's NetzDG Law. *Transatlantic Working Group*. Retrieved from: https://www.ivir.nl/publicaties/download/NetzDG_Tworek_Leerssen_April_2019.pdf.

World Health Organisation. (2020, 11 December). Call for Action: Managing the Infodemic. Retreived from: https://www.who.int/news/item/11-12-2020-call-for-action-managing-the-infodemic.

York, J.C. (2021). *Silicon Values. The Future of Free Speech Under Surveillance Capitalism*. Verso: London.

Zuckerberg, M. (2019, 30 March). Opinion. Mark Zuckerberg: The Internet needs new rules. Let's start in these four areas. *Washington Post*. Retrievecd from: https://www.washingtonpost.com/opinions/mark-zuckerberg-the-internet-needs-new-rules-lets-start-in-these-four-areas/2019/03/29/9e6f0504-521a-11e9-a3f7-78b7525a8d5f_story.html.

CHAPTER 8

Insights from South Asia: a Case of 'Post-truth' Electoral Discourse in Pakistan

Anam Kuraishi

This chapter proposes a shift in thinking about post-truth from conflating post-truth to an 'era' and/or 'politics' to associating post-truth with discourse. This is done by contextualizing post-truth within the ambit of rhetoric and ideology and thus distancing the discussion from two oft-recurring flaws in the mainstream debate: fixation on 'post' as *something after* truth and overemphasis on misinformation. When discussing post-truth in relation to discourse, the focus lies in examining the specific constituents and attributes of discourse which render it 'post-truth'. This is achieved by developing a guide to operationalise the term 'post-truth' in order to identify such discourse. The ensuing toolkit is then applied to the case study of Pakistan's electoral discourse, focusing specifically on newspapers. Such an operationalisation provides us with the tools to evaluate and categorize political discourses, claims, narratives, and accounts as 'post-truth'.

The chapter is based on a qualitative evaluation and analysis of a set of 1,205 news articles and opinion pieces that covered major incidents, trending political issues and election speeches selected from the three major English-language newspapers (Dawn, The News International, The Express Tribune) circulated in Pakistan over the time frame of three National election years (2008, 2013 and 2018). Drawing upon the logics of critical explanation approach (Glynos and Howarth 2007), specifically the phantasmatic logic, and Lacanian psychoanalytical elements of *lack* and fantasy, intertwined with *emotionality*, a categorization of political accounts in newspaper reporting is developed to evaluate their 'post-truth' status. It is suggested that the mechanism of categorizing accounts as 'post-truth' is a three-step process: (i) identification of a specific *lack* the constructed narrative offers to fulfil; (ii) identification of a specific *fantasy* which structures and addresses a specific lack; and (iii) identification of the *emotional* element in the texts used to construct the narrative; all three steps intertwine to depict a *phantasmatic logic* which renders a discourse 'post-truth'.

 | DOI:10.1163/9789004547261_010

1 (Re)Defining 'Post-truth'

Post-truth has gained traction in the recent political landscape with phrases such as 'post-truth era' or 'post-truth politics' becoming a norm to describe the political sphere. Most commonly, post-truth is associated with fake news and misinformation: the truthfulness of an account is claimed to now be secondary and is disregarded, allegedly giving way to the use of emotional appeals to influence perceptions of individuals about how reality is experienced and to attempts at denying 'factual truth'. Therefore, it only seems befitting that post-truth has been prominently discussed in light of Brexit claims of 'taking back control' or Trump's claims of 'draining the swamp'. The emphasis is largely placed on emotional appeals used to influence perceptions of individuals about how reality is experienced.

The majority of the current scholarship on post-truth falls within the ambit of mainstream media conceptualisation thereof, which is focused on the ideological supremacy of emotionality at the expense of 'truth'. There are three attributes which are usually highlighted in association with post-truth in the literature – the erosion of truth in the political domain (Arendt 1972; Bagginni 2017; Higgins 2016; Hyvönen 2018; Keyes 2004), replacement of truth with emotional appeals to influence public opinion (d'Ancona 2017; Davies 2016; McComiskey 2017; McIntyre 2018; Orwell 2017), and the rise of misinformation at the expense of belief in objective truthful information (Barrera et al. 2020; Bilgin 2017; Black 1982; Davis 2017; Flynn et al. 2017; Frankfurt 2009; Lewandowsky et al. 2012; Nyhan and Reifler 2010; Nyhan and Reifler 2015; Kuklinski et al. 2000; Seifert 2017; Shin et al. 2018). These frameworks offer insights into the elements of information and how it is processed and experienced in today's political realm. However, such a discussion on 'post-truth' is restrictive in interpretating the recent developments in the political domain.

It is proposed that thinking about post-truth ought to shift from conflating post-truth to an 'era' and 'politics' to associating it with discourse. That should be achieved by way of by contextualizing post-truth within the ambit of rhetoric and ideology instead. Likewise, 'post' must be disassociated from being *something after* truth while the overemphasis on misinformation has to be reduced as well. When discussing post-truth in relation to discourse, the focus lies in examining the specific constituents of discourse which render it 'post-truth'. In doing so, Lacanian psychoanalysis is a useful approach to draw from and, particularly, his conceptualization of *lack* and *fantasy*, leading to a logic of critical explanation, i.e., phantasmatic logic (Glynos and Howarth 2007). This helps develop a guide for operationalising the term 'post-truth' in order to identify 'post-truth' discourse. To illustrate the applicability of this framework, a case study of Pakistan's electoral discourse in newspapers is carried out, determining the extent of 'post-truth' electoral discourse in the country. Such an operationalisation provides us with the tools to evaluate and categorize political discourses, claims, narratives, and accounts as 'post-truth'.

Single case studies in comparative politics have been resurging in the recent years. Pepinsky (2019) argues that single case studies have evolved over time, with an emphasis on hypothesis testing and research design. However, East and South Asia remain underrepresented in such research (Pepinsky 2019, 198). Moreover, in terms of post-truth, the region has thus far been largely devoid of analysis, there being several potential reasons for such an occlusion, not least, the innate assumption that post-truth is not applicable in a non-western context or the choice of the (primarily western) researchers to conduct an empirical study in the contexts they are familiar with and. Therefore, the decision to use Pakistan as a case study was to dissociate from the innate assumption that post-truth cannot be studied in a non-western context as well as to empirically illustrate the operational definition of 'post-truth' presented in this study, which can be replicated further in various contexts.

Pakistan has had a history of military coups with instances of democratically elected governments. Such a fractured political history has shaped public opinion on the forms of regime type as well as informing identities surrounding political narratives on various issues. There are three reasons as to why the electoral discourse is chosen as the starting point for empirically illustrating 'post-truth' political accounts and narratives: (i) to provide insight on how a political narrative surrounding election is shaped in Pakistan, and whether the traits attributed to the terminology 'post-truth' are applicable to newspaper articles covering electoral discourse in Pakistan; (ii) to provide additional insight on whether the chequered political history of Pakistan has implication on the process of creating a political narrative, and subsequently how that ties in with the construction of 'post-truth' electoral narratives; and (iii) to evaluate the content of the political narratives in three different election years in order to determine if electoral discourse evolves over time and how discourses are structured at different times.

2 Lacanian Psychoanalysis and Critical Logics of Explanation

Discourse theory focuses on explaining and understanding the conditions under which discourses are created and contested. There is not a single prescriptive methodological approach for conducting theoretically informed research on discourse. Instead, there are various guiding principles identified by scholars which inform the conduct of discourse-theoretical research (Glynos and Howarth 2007; Howarth 2000). The application of discourse theory to empirical research topics is contingent upon the particular problem researchers are focusing on and modulating and articulating their concepts accordingly. The research strategy employed in association with discourse theory for this chapter is in light of the logics of critical explanation approach (Glynos and Howarth 2007), in particular the phantasmatic logic in addition to drawing on the Lacanian psychoanalytic theory (Lacan 1977). Particularly emphasis is put here on the concepts of *lack* and *fantasy*. This approach, it is suggested, provides us with the elements needed to evaluate the patterns in political discourse and the creation of a political narrative, which can be categorized as 'post-truth'.

The logics of critical explanation (LCE) is a conceptual framework witin the field of post-structuralist discourse theory developed by Glynos and Howarth (2007) to conduct empirical research on questions pertaining to social sciences. In particular, the logics of critical explanation framework is meant to be used as a tool to explain and critique the practices and regimes that constitute concrete objects of analysis – logic is the basic unit of explanation through which a problematized phenomenon in the social sciences can be explained (Glynos and Howarth 2007, 132–133). The process of explanation involves mobilization of three separate yet interrelated types of logic – social, political, and phantasmatic – which build upon and supplement Laclau and Mouffe's conceptual framework of discourse theory (Laclau and Mouffe 2014). Social logic refers to certain social practices and how they become sedimented through a socialization process. Subjects adhere to certain rules and practices to give meaning to the world around them and subsequently shaping how they perceive the reality (Glynos and Howarth 2007, 137). Political logic, meanwhile, focuses on the emergence of regimes – the creation of a discourse or an institution which sediments a social practice, i.e., institutionalization of the social. Phantasmatic logic builds upon this notion of continuity and contestation of social practices and regimes. It explains why a particular social practice or regime – a hegemonic articulation – is able to maintain its 'grip' over the subjects (Glynos and Howarth 2007, 145). It focuses on the ability (or the inability) of a specific discursive formation to resist a change to a certain social practice and maintain its prominence.

The operation of a fantasmatic logic is premised upon the Lacanian concept of the fantasy which plays a key role in 'filling up' or 'completing' the void in the subject and the central structure of social relations by bringing a closure (Glynos and Howarth 2007, 146). Fantasies in this sense provide meaning to the world for the subject, giving consistency to what the subject perceives as reality; fantasies are 'the support that gives consistency to what we call "reality"' (Žižek 1989, 44). Fantasy operates to bring forth a closure for a subject's identity, "promising fullness once an obstacle is overcome – the beatific dimension of fantasy – or which foretells of disaster if the obstacle proves insurmountable, which might be termed as the horrific dimension of fantasy" (Glynos and Howarth 2007, 147). Hence, fantasy can be seen as aspirational.

The elements of *lack* and *fantasy* are crucial components required for categorizing political accounts as 'post-truth'. A brief description of the Lacanian approach to subjectivity and fantasy is needed to understand the hegemonization of a specific discourse to be accepted as being truthful. Lacan's conception of the subject is shaped by the Freudian notion of *Spaltung* (splitting) – the subject is structured around a radical split and constantly aims to overcome this split, or lack (see Glynos and Stavrakakis 2008). For Lacan, the concept of *lack* is understood as a lack of jouissance (enjoyment), and it is this reclamation of jouissance which is offered through resort to *fantasy* or, to put it more aptly, the phantasmatic vision of recapturing this supposedly lost fullness (Glynos and Stavrakakis 2008). The *lack*, therefore, refers to the deficiency presumed by the individual constituting their self-identity and it is related to the aspiration (or desire) to achieve a joyful fulfilment in their lives. The aspirational belief that individuals possess about achieving something resonates with this notion of the *lack* and is also reflected in their perception and belief about a particular discourse. A *fantasy* is related to the identification of a subject (individual) – *fantasy* serves as a way of trying to provide meaning and resolve the *lack* one associates with their own identity. A *fantasy*, therefore, should be understood as a schema linking the subject to a socio-political reality with a reference to the object-cause of desire (Glynos and Stavrakakis 2008) – imitating a fantasmatic ideal. The *lack* requires that one identifies with socially acceptable traits of identification found, which are aimed at re-imagining and re-instituting an identity till the individual reaches the point where the *lack* in them diminishes.

This chapter discusses a specific example of one type of political narrative that can be categorized as 'post-truth' – discourse on democracy and democratization in Paskistan. A qualitative evaluation and analysis of a set of 1,205 news articles and opinion pieces that covered major incidents, trending political issues and election speeches selected from the three major English newspapers (Dawn, The News International, The Express Tribune) that had been circulating in Pakistan over the time frame of three National electoral years (2008, 2013 and 2018) is conducted. It is suggested that the mechanism of categorizing accounts as 'post-truth' is a three-step process:

1. Identification of a specific lack the constructed narrative offers to fulfil;

2. Identification of a specific fantasy which structures and addresses a specific lack;
3. Identification of the emotional element in the texts used to construct the narrative.

All these conditions are necessary but not sufficient on their own. The intertwining of these three steps indicates the operation of the phantasmatic logic underpinning the narrative, and its capacity to maintain its 'grip' on the audience, rendering a narrative 'post-truth'. Hence, to further pin down the specific post-truth qualities, it is proposed that a 'post-truth' narrative adapts to a particular context and evokes a fantasy structured around a specific type of available knowledge and information, structuring and addressing a lack in an individual. Accordingly, a specific ideal identity is created and informed, which the individual aspires of attaining. It is inclusive of phantasmatic content, the aspirational identity which is inclusive of the elements of yearning and belonging guised in emotionality. The yearning for a positive outlook and the shared experiences are what constitutes the lack. A 'post-truth' narrative is such that is able to construct and reconstruct identities in its aim of presenting specific versions of reality and redemption. The peculiarity of a post-truth lack is dependent on the fantasy – the fantasy advocates certain identities and refers to the lack in a particular manner aiming to evoke a sense of yearning and belonging to the audience.

3 The Lack of Aspired Life and the Fantasy of Socio-political Development

The *lack* present in the discourse on democracy refers to the degeneration of society socially, economically, and politically which has resulted in the hampering of development and progress along with a low quality of life for citizens. It refers to what is presented as the degenerate condition of Pakistan and the impediment of advancement as a result of incompetent rulers and the garb of power at the expense of the neglect of the welfare of citizens – the unfulfilled desire of an individual to have an improved lifestyle and opportunities for progress. It is the absence of an opportunity to improve their livelihood – a void which keeps the individual unfulfilled, dissatisfied, and frustrated. This *lack* is associated with the form of governance in place in the country and is presumably present in the individual as a result of surroundings and the experiences the individual is faced with. The *lack* also refers to the freedoms – social, political, and economic which are unattainable in the status quo. The absence of a means of improving livelihood impedes the individual in finding closure for their identity. It is this condition of the individual that is being targeted with the rhetoric of change – the promise to eradicate this dissatisfaction and frustration by offering a means whereby the individual can fill the void they associate with their identity. The void referring to the aspirational desire of an individual to exercise their right of choice in electing a leadership which will deliver on its promises, cater to the well-being of the citizen, along with providing the means to be able to prosper.

The *lack* in the narrative on democracy refers to the current misery, low quality of life of the public, the degeneration of society socially, economically, and politically, along with the failure of the incumbent government to prioritise citizens over their own personal gains. Here are some examples of how the lack is being framed:

> President Zardari and his 'brother' Nawaz Sharif had completed their tenures and they both failed to rid the masses of poverty and unemployment. Peace, stability and a corruption-free society could only be established through justice, but the former rulers promoted the 'thana and patwari culture' for their vested interests instead of serving the masses.[1]
>
> The rulers had also miserably failed to launch any welfare project for poverty-stricken masses. The incumbent governments had totally failed to deal with militancy and restore peace.[2]
>
> We'll bring back the looted money of these corrupt and greedy rulers from abroad and spend it for the welfare of the people, who are even deprived of health and education services due to […] corruption. [3]

It is the dismantling of this status quo which the appeal of change geared towards – the promise of a new and better tomorrow that the *fantasy* of change guised in democracy is aimed at delivering – "to make Pakistan great again."[4] But it is also the inability to entirely achieve the desiring aspect of a *fantasy* fully – the incompleteness of the *lack* which makes an individual to continue aspiring for a fulfilment.

The *fantasy* underpinning this *lack* is the rhetoric of change guised under the ideals of democracy and democratization – an alternate form of system which can bring forth an affirmative progression for the individual. The *fantasy*

1 2013 Main News, "PTI To Unite People in New Pakistan: Imran", Dawn, 7 May 2013.
2 2013 Main News, "PPP, N Supporting Terrorists: Imran", The Nation, 5 March 2013.
3 2013 Main News, "If Elected to Power: We'll Confiscate Assets of Nawaz, Zardari: Imran," Dawn, 6 March 2013.
4 2018 Opinion Piece, Cyril Almeida, "Noisy Politics", Dawn, 29 April 2018.

works in two ways; a) it creates a sense of multiple yearnings: choosing freely and independently, redemption from the degenerate status quo, regaining control to serve one's interests etc.; and b) it emphasizes the sense of belonging – a shared experience. The *fantasy* of change during an electoral campaign entails the promise of an improvement on the status quo of the country – the progression of Pakistan socially, economically, and politically. It is the promise of bringing forth honest leadership which will serve and deliver for the welfare of the citizens. A narrative which is able to provide a meaningful account and is relayed as being plausible is likely to be perceived as being true. The acceptance of such a narrative to be truthful and acting upon it through the exercise of one's vote offers *jouissance* – the reclamation of enjoyment – the fulfilment of a desire to be better off if honest leadership is elected is able to provide closure to one's identity which one yearns. It is here that the role of *emotionality* becomes pertinent – the narrative becomes an alternate form of evidence. *Emotionality* evokes a sense of belonging – a yearning for improvement and a shared identity of experiencing the status quo. The *fantasy* is able to construct a social identity in response to the current identity of the individual – a promise for redemption.

The above, then, necessitates the application of a phantasmatic logic as an analytical framework. The phantasmatic logic is intended to explain why a particular social practice is able to maintain its 'grip' over the subject (Glynos and Howarth 2007, 145) – the ability of a discourse to maintain its prominence. The phantasmatic narrative which "offers fullness-to-come once an obstacle is overcome" (Glynos and Howarth 2007, 147) – the beatific *fantasy* in the case of democracy – represents the rhetoric of change guised in the principles of democracy and democratization. That ranges from restoring democracy to consolidating democracy, and is intended to fill the void present in the absence of a democratic governance system that would positively impact the livelihood of an individual. The obstacle in this case is both external and internal. External obstacles include the interference of military in politics, corrupt leaders, and their incompetence, which have resulted in the degradation of the livelihood of the individual. Internal obstacles include the conservative ideals intertwined with religious values – ant these are in contrast to some of the values that democracy espouses. It is indeed the presence of these obstacles that maintain the closure or hinder the fulfilment of the *lack* – the horrific dimension of *fantasy*.

On the other hand, the discourse on democracy highlights the presence of other forms of identities in parallel – the opposing identities which are informed by other discourses. These identities refer to Islamist discourse on democracy being antithesis to Islam, or religious conservatism which is in stark opposition to the ideals propagated by the *fantasy* of democracy. It is worth noting here how the horrific dimension of *fantasy* plays out. The horrific dimension refers to the sustenance of a status quo with the continuation of the *lack* if the 'obstacle proves insurmountable' (Glynos and Howarth 2007, 147). In the discourse on democracy, the obstacle refers to the inability of the 'change' advocated through a democracy being unable to take root, either because of the 'internal obstacle' (enemy within) (Glynos and Howarth 2007, 150) which in this case refers to the internalisation of conservative values that will need to be displaced if a democratic rule is to prevail. For example, the mobility of women in public spaces, a freedom which comes through the implementation of a democratic rule but stands at odds with a society's conservative ideals which are intertwined with religious values whereby a woman's position is within the household or requires a certain level of purdah (veil) in terms of segregation between men and women. In other instances, the *fantasy* does not take root because the individual fails to resonate with the ideals and identity which is informed by the discourse or the discourse on its own fails to appeal to the targeted individual, leading to the sustenance of the status quo. So, whereas in one instance a positive identity is advocated through a *fantasy*, in similar vein the *lack* of that identity is also indicated if the said *fantasy* is unable to foster a sense of belonging and yearning and is therefore unable to take root in society.

4 The Jouissance of the Fantasy of Change Disguised as Democracy

Democracy acts as promise for citizens to honour their vote and their right of choice – to give them the liberty and freedom to exercise their right of choice and the powers that come with it – the power of holding elected representative leaders accountable. The phantasmatic marker deployed in appealing to the audience is accountability – holding the incumbent government responsible and making them answerable to the citizens whose vote was dishonoured. The selection of a phantasmatic marker is based on the rhetoric associated with the *fantasy* that is propagated in newspaper reporting on a selected topic. The phantasmatic label acts as the tool through which a *fantasy* is able to ground itself. The phantasmatic marker can be further categorised into removal of incompetent leaders, eradication of dynastic and patronage politics, free and fair elections, removal of corruption, removal of military interference. The rhetoric of change therefore is meant to replace the incompetent political leaders who have looted and plundered the country and not delivered on policies meant to generate progression. The *fantasy* of change was predominantly the narrative espoused by Imran Khan, the leader of the PTI who blamed the incumbent governments of Pakistan Muslim League-Nawaz (PML-N) and the Pakistan People's Party (PPP) "equally responsible for difficulties the people suffered."[5]

Pakistan has had a chequered history of types of governments – it has witnessed four military coups since its inception with periods of democratic governments in between. However, there has always been the struggle for democratic consolidation in practice, and political parties have time and again used the rhetoric of being the harbinger of democracy and democratic principles throughout their electoral campaigns. Pakistan's political history can be succinctly summarized as follows:

"Pakistan's journey towards democracy has been rocky would be an understatement. The country's experience in embracing democratic values and culture still remains a long way off. There are still those dark forces that do not believe in democracy. Either because they suffer from an inferiority complex and feel that we as a people are not fit for it. Or for selfish reasons realise that they would be losers if democracy were to replace dictatorship or military rule."[6]

The struggle for democracy has been a common theme present in the political discourse surrounding elections in all the electoral years analysed. The overarching *fantasy* refers to the constitution of government which is elected by the public – the honouring of a citizen's right to elect representatives they want to govern the country along with a sincere leadership which delivers on its promises and works for providing welfare to its citizens. The *fantasy* of redemption structures a reality whereby the individual's rights will be honoured and respected – where the individual will have the means to thrive. However, the rhetoric surrounding the restoration of democracy has varied in the three electoral years. The *lack* that the *fantasy* of democratization is aimed at addressing refers to the curtailment of the citizen's right to vote and the dishonouring of what their vote stands for.

Prior to the 2008 election, it "was the first time in 60 years of Pakistan's life that an army chief was elected president"[7] between 2002 and 2008. In October 2007, "Pervez Musharraf won a one-sided election for another five-year term

5 2013 Main News, "PTI To Field Candidates in All Constituencies Of KP", Dawn, 3 March 2013.

6 2018 Opinion Piece English, Talat Masood, "How Should I Vote?", The Express Tribune, 20 June 2018.

7 2008 Main News English, "Musharraf Steals the Show, But Victory Hangs on Court", Dawn, 7 October 2007.

from a truncated parliamentary electoral college amid boycotts and protests",[8] a day which has been labelled as a "black day"[9] saw protests staged by opposition parties demanding that General Pervez Musharraf[10] resign "before the people of Pakistan throw him out of power".[11] The participants in the rallies chanted slogans in favour of Benazir Bhutto and Nawaz Sharif[12] – "Benazir Bhutto and Nawaz Sharif … will return with the power of the people of this country," Mr Rabbani said.[13] In November 2007, General Musharraf enacted an emergency rule[14] where "he staged a coup against his own office as president and, acting as the army chief, promulgated a state of emergency, imposed martial law, suspended the Constitution, sacked and placed under house arrest judges who had held their heads high and declined to bend the law to suit his whims and wishes."[15] The *lack* that the restoration of democracy narrative is addressing is the absence of consolidation of democracy and the degeneration of democracy in Pakistan and an imposition of martial law which infringes upon the right of people to choose representatives they would want to govern their country – it is the absence of the right to exercise their choice of an elected representative and civilian rule – socio-political freedoms came across as the phantasmatic markers in advocating a *fantasy* of what a democracy had to offer to the citizens. The *fantasy* therefore is to offer means whereby the civilian rule can flourish and separation of powers and interference of the military in the political realm is made possible.

The staging of protests against the dictatorial rule and the "demand for restoring democracy in countries like Pakistan"[16] and "expressing solidarity with the legal fraternity, the media and for the independence of the judiciary"[17] along with using appeals like "PPP's chairperson Benazir Bhutto was the

8 2008 Main News English, "Musharraf Steals the Show, But Victory Hangs on Court", Dawn, 7 October 2007.

9 2008 Main News English, "Musharraf Steals the Show, But Victory Hangs on Court", Dawn, 7 October 2007.

10 General Pervez Musharraf Is A Retired General Who Became the President of Pakistan Between 2001 Until 2008.

11 2008 Main News English, "KARACHI: MMA Joins ARD Protest Rally", Dawn, 27 March 2007.

12 2008 Main News English, "KARACHI: MMA Joins ARD Protest Rally ", Dawn, 27 March 2007.

13 2008 Main News English, "KARACHI: Struggle to Continue till Removal of Musharraf: ARD Rally Vows …", Dawn, 14 April 2007.

14 2008 Main News English, "Provisional Constitutional Order", Dawn, 04 November 2007.

15 2008 Opinion Piece English, "Op-Ed Perceptions and Misperceptions", Dawn, 17 February 2008.

16 2008 Main News English, "Benazir Says She Will Return to Pakistan By Early November", Dawn, 5 February 2007.

17 2008 Main News English, "Karachi: MMA Joins ARD Protest Rally", Dawn, 27 March 2007.

last hope for restoration of democracy and civil rule in the country"[18] are all examples of how *emotionality* in form of an act is used as a tool to appeal to the public. The aim of these acts is to evoke a sense of belonging amongst the public of the similar *lack* they are facing and act as a means whereby they can persuade the public to believe in the gains that a democratic civilian will bring – freedom of choice which is every citizen's right and has been revoked by the imposition of a military dictatorship.

In 2013, the *fantasy* of restoring democracy was constructed around the ideals of will of the people:

> True democracy stands for a government in which the people's will prevails. It stands for justice, equality, accountability, and transparency. It ensures the rule of law and supremacy of the constitution. It paves the way for good governance, which in return gives the people peace, progress, and prosperity. If a system of government lacks these basic features, then it is not a democracy, but a mockery of democracy.[19]

The *lack* being addressed here is the waning of democratic rule and the lack of accountability for elected leaders – it is the non-fulfilment of the promises that elected politicians made during their campaigning about development and progression of the society socially and economically. The phantasmatic marker for restoring democracy in 2013 reflected the appeal for making leaders accountable for their actions using emotional tactics of referring to the sacrifices made the political parties for the struggle for democracy and alluding to the glorious foundation of creating Pakistan a democratic state. The emphasis is placed on the supremacy of a civilian rule over a dictatorship for the country's survival[20] – "the continuity of democracy is the right of the people and Pakistan cannot afford any unpredictability right now [...] elements trying to derail democracy" by "followers of dictatorship" and said they were placing the country's future at stake.[21] The underlying marker for the success of the

18 2008 Main News English, "CEC Takes Notice of Ad Campaign Against Benazir", Dawn, 17 November 2007.

19 2013 Opinion Piece English, Ahmad Noor Waziri, "In the Name of Democracy?", The Nation, 9 April 2013.

20 2013 Main News English 2013, "PML-N Pockets Baluchistan Political Gurus", The Nation, 6 March 2013.

21 2013 Main News English,"Qadris Long: March, PML N-Leaves 'The Matter' To Federal Government", The Express Tribune, 3 January 2013.

civilian rule for consolidating democracy is holding them accountable for their actions.

> The country has drifted deeper into an abysmal political chaos and economic uncertainty. The common man's life could not be more miserable with uncontrolled food and energy shortages, unabated violence and countrywide lawlessness.[22]

Political leaders mobilised the audience with referring to the sacrifices their parties have made for the struggle for democratising Pakistan, using partisanship as an emotional tactic to evoke a sense of belonging amongst the audience for who can deliver 'true democracy' – "The murderers of Quaid-i-Azam and Benazir Bhutto (BB) Shaheed now want to eliminate us as well (but) I don't care for my life. The world knows that the PPP has laid down lives for democracy."[23] Along similar lines, the act of staging a "march for democracy"[24] and referring back to "restoring Jinnah's true democracy"[25] are meant to advocate for a better governing system by glorifying the ideals of a democratic state as envisioned by Jinnah – the founding father of Pakistan. The ideals refer to an inclusive and impartial government, freedom of law, religion, and equality for all.[26] It is through reminiscing of the good days that politicians evoked a sense of community amongst the audience – appealing to them for the glory days by making the incumbent leaders accountable and exercising their right to vote cautiously.

The 2013 discourse surrounding democracy referred to the incumbent civilian rule as dictatorial rule – it was primarily the dynastic politics followed by military interference which dominated the narrative of regaining control. There is a contrast from the 2008 narrative which focused solely on removing a military dictator from politics and restoring democracy. However, a reference to the military interference remains a constant reminder to the dark days

22 2013 Opinion Piece English, Shamshad Ahmed, "Our Revengeful Democracy", The Nation, 9 April 2013.

23 2013 Main News English, "Bilawal Kicks Off Campaign with Video Message", Dawn, 24 April 2013.

24 2013 Main News English, "Qadri's Long March Departs from Lahore", Dawn, 13 January 2013.

25 2013 Main News English, "Tahirul Qadri Set to Address Rally in Karachi", The Express Tribune, 1 January 2013.

26 Mr. Jinnah's Presidential Address to The Constituent Assembly of Pakistan, August 11, 1947: http://www.pakistani.org/pakistan/legislation/constituent_address_11aug1947.html.

where military intervention risks the lack of freedom and right of choice, the reasons for which can be summarised as follows:

> In reality, the fact remains that the military remains a powerful political actor in Pakistan … the fact that the military has chosen not to directly interfere in politics in the past decade does not imply that it has not played a role behind the scenes.[27]
>
> The country cannot afford any further dictatorship which is dangerous for its survival.[28]

The informal power exercised by the military is a constant threat to civilian rule in Pakistan which also feeds into the recurring *lack* associated with the individual. The threat of military interference is compounded by the incapacity of civilian rulers to establish their own legitimacy amidst meagre governance and allegations of corruption. Therefore, the rhetoric of change surrounding the 2013 elections was built on the notion of removing the 'corrupt politicians' and 'circus lions' and referring to 'the day of accountability'[29] with the appeal of asking individuals to elect a new leadership which can bring about a change in the country and put it on the road of progression.

> The sad part of the story is that Pakistanis have been betrayed several times in the name of democracy. Dictators have intervened on the pretext of cleaning the mess created by the politicians to let democracy flourish. While the politicians seek power by making big claims of serving the cause of democracy. In reality, they have served no one – neither people nor democracy.[30]

Most of the rhetoric emphasised on the word 'change' and a 'new Pakistan' as can be seen from the examples below.

> People want change. They want a change in the system not just a change of faces.[31]

27 2018 Opinion Piece, Hassan Javid, "Moving Beyond the Dawn Leaks", The Nation, 14 May 2017.

28 2013 Main News, "PML-N Pockets Baluchistan Political Gurus", The Nation, 6 March 2013.

29 2013 Main News, "Imran Calls Upon Bombers to Stop Attacks", Dawn, 23 April 2013.

30 2013 Opinion Piece, Ahmad Noor Waziri, "In the Name of Democracy?", The Nation, 9 April 2013.

31 2013 Main News, "Tribesmen Embrace Historic Polls," Dawn, 5 May 2013.

> May 11th will be the end of obsolete system and sun of new Pakistan will rise … start a golden era … completely change the Pakistan according to the dreams of the masses.[32]
>
> "Seize the moment of change." "You, the people, have to decide if you want to carry on as it is or bring about a change".[33]

Democracy – or rather 'true' democracy is a rallying point for most of the politicians, each bidding that only they can establish it by being in power and the citizenry need to make the correct choice. The rhetoric of change was prominently advocated by the Pakistan Tehreek-e-Insaaf (PTI) which gained traction in 2013 and continued to dominate the political sphere in the subsequent electoral cycle as well. The word change was referred in conjunction with justice (*inquilab*) for the people who have suffered at the hands of the corrupt leadership – the removal of incompetent incumbent leadership referring to mostly the two dynastic political parties, the Pakistan Muslim League-Nawaz (PML-N) and the Pakistan People's Party (PPP). The emotional appeal surfaced in the usage of specific words and phrases such as "inquilab"[34] – justice; "million-man march" and "Safar-e-Inqilab-e-Pakistan"[35] [36] – reference to the strength of people's choice and the journey towards justice and accountability; "tsunami"[37] – reference to the revolution which will take place with swarms of people voting for a new leadership, sweeping the incompetent incumbent leaders out of power; "waiting for upcoming May 11 for the last 17 years"[38] – reference to the time period that the incompetency and corruption has continued for, causing the degradation of the society, and hampering the livelihood of the citizenry and emphasising on the need of the hour to take control back in their hands by removing insincere leadership; "Naya Pakistan – One Pakistan"[39] – reference

32 2013 Main News, "Imran Sees Battle Between 'Noon' And 'Junoon'," The Nation, 30 April 2013.

33 2013 Main News, "Imran Bounces Back, Keeps Hope Going," The Nation, 8 May 2013.

34 2013 Main News, "Thousands Gear Up for Inquilab At Jinnah Ground", The Express Tribune, 1 January 2013.

35 2013 Main News "Tahirul Qadri Set to Address Rally in Karachi," The Express Tribune, 1 January 2013.

36 2013 Main News "Tahirul Qadri Set to Address Rally in Karachi," The Express Tribune, 1 January 2013.

37 2013 Main News, "From Geek to Chic," Dawn, 5 May 2013.

38 2013 Main News, "PTI Out to Hunt Down Lion," The Nation, 8 May 2013.

39 2018 Opinion Piece, Mohsin Raza Malik, "A State of Political Entropy," The Nation, 9 May 2018.

to a new socio-political order based on equality, social justice, accountability and pluralism.

Why the usage of 'change' and its association with a 'new Pakistan'? A new Pakistan can only be formed if there is a removal of corrupt politicians and dynastic families which have ruled the country for decades replicating an autocratic rule, stripping people of their right to hold leaders accountable and plunging the country into an abyss. The failure on part of the incompetent incumbent leadership to deliver on its promises forms the premise of introducing 'change'. The word change here is referring to two aspects: one, a change in the leadership, and two, a change in the outlook of the country; both changes referring to the creation of a prosperous country which leads to the improved livelihood of the citizenry.

> Our struggle is not just 'Go Nawaz Go' … it is 'Go System Go', Imran said, adding that the current system of the country thrived on corruption and nepotism. "Everybody will be equal in Naya Pakistan".[40]
>
> Ads for Pakistan Tehreek-e-Insaf (PTI) Imran Khan offer voters a "new Pakistan" with his PTI party symbol, a cricket bat, swiping away corruption and propelling the country into the future.[41]

The rhetoric of accountability and rule of law formed the crux for the rhetoric of change – the reiteration of the strength of an individual's vote, the emphasis placed on placing power in the hands of the common people, and the references to a new socio-political order acted as markers of emotional appeal. Additionally, two developments in the political sphere fed into the rhetoric of change surrounding the 2013 and 2018 electoral years: (i) the emphasis placed on the role of youth in elections; and (ii) the rise of fringe parties. With regards to the former, the appeal surrounding the future of the country, prosperity and livelihood of the citizens was targeting the future generation – people who will continue suffering if a change is not implemented. The fact that "over 25 million youth"[42] would exercise their right to vote in the upcoming electoral cycles, acted as a compelling reason to target this audience with the rhetoric of change and what a new Pakistan could mean for them. The latter aspect, the

40 2018 Main News, "Azadi March Imran Khan Claims Breaking", The Express Tribune, 27 September 2014.

41 2013 Main News, "Election 2013: Political Parties Channel Millions into Ads", The Express Tribune, 3 May 2013.

42 2013 Opinion Piece English, Imran Ali, "Opening Doors Wider", The Express Tribune, 2 May 2013.

rise of fringe parties put forth the idea of 'change' in perspective – the emphasis placed on the power of people's right of choice and the demand for the supremacy of rule of law made the incompetency and corruption of previous leadership part of the political conversation. Whereas PTI was the harbinger of introducing the idea of political change, the Pakistan Awami Tehreek (PAT) and Tehreek-e-Labaik (TLP) bolstered the power of *dharnas* (a non-violent sit-in protest) to reinforce the power the citizens had in regaining control by exercising their right to vote for sincere leadership which can put the country on the road to transformation and progress.

Prior to the 2018 election, the *fantasy* of democratisation referred to both restoring and consolidating democracy as the "second consecutive transition from one elected government to another was being made, a significant milestone indeed in the country's rocky democratic political journey."[43] The phantasmatic marker underpinning the *fantasy* was accountability of elected leaders and their disregard of the welfare of the public – the absence of fulfilling promises for social and economic progress of the country. The *lack* being targeted was the degradation of the social and economic aspects of the citizen's quality of life. The fantasmatic ideal of democratization alluded to the progress of the society under sincere leadership and holding incumbent politicians accountable for the degeneration of the society by bringing them to justice:

> democracy and autocracy will fight it out in Islamabad to determine the destiny of a new Pakistan ... no one will gift you freedom on a silver platter. You'll have to snatch it through a sustained struggle.[44]

It was successive *dharnas* by the PTI over the results of the 2013 elections[45] which set the premise of struggling for democratization as the basis of electoral campaigning for 2018. It is the promise of a free and democratic state that is being espoused by the chairperson of PTI, Imran Khan who appeals to the public by advocating for a 'sustained struggle' and illuminating that the role of elected leaders is not to attain power for personal gains but to deliver on promises and serve the citizens. The emphasis was placed on "rule of law

43 2018 Opinion Piece English, Zahid Hussain, "Horses of The Same Stock", Dawn, 13 June 2018.

44 2018 Main News English, "PTI Kicks Off Azadi March", Dawn, 15 August 2014.

45 2018 Opinion Piece English, Hassan Javid, "Shrinking Democratic Space", The Nation, 15 April 2018.

and accountability as being two important pillars of a liberal democracy",[46] an aspect which had been disregarded by PML-N government which was elected in 2013 – "the PML-N's disregard for real political work will darken democracy's prospects"[47] because of "a lack of crucially required reforms resulting in waning public faith in democracy."[48] The fantasmatic ideal propagated was based on the progression of the society under a leadership which prioritises the citizens over personal gains – the aim of Imran Khan's rhetoric was to induce a sense of belonging amongst the audience regarding the loss of the meaning of their vote – the dishonouring of the citizen's vote for the incumbent government which failed its citizens in delivering upon its promises. The appeal for the audience focused on regaining control over their choices and holding the government accountable for its disregard of the promises to advance public welfare.

In a nutshell, the *fantasy* of 'change' and the promise underlying democracy and democratization is focused on advocating for honouring the citizen's right to choose an elected representative and hold them accountable for their actions which is the basis of a democratic state. The alternative form of reality is constructed through emotional appeals of correcting the system and the disadvantaged status quo which have caused suffering of the citizenry. In parts, the alternate identity indicated is the curtailment of freedoms if military interferes in the political sphere and the conservative values which are at opposition to the democratic freedoms. Therefore, the multiplicity of antagonistic discourses and identities surrounding the theme of democracy are illustrated. *Emotionality* was used in various manners from staging successive peaceful sit-in protests and appealing to audience using partisanship, references to the glorious history of creating Pakistan in the name of democracy by illuminating the loss the citizenry were suffering from at the hands of a non-democratic regime. The *lack* which was being addressed referred to the degeneration of society and the suffering of the citizen which was caused by the waning of democratic principles and the interference of military institution in the political domain curtailing and disregarding the freedom of choice for citizens. The phantasmatic logic underpinning this narrative is the vision of a prosperous country under sincere leadership which honours the individual's right of choice and serves for their welfare. Therefore, the discourse surrounding the

46 2018 Opinion Piece English, Zahid Hussain, "Democratic Accountability", Dawn, 6 June 2018.

47 2018 Opinion Piece English, I.A. Rehman, "Democracy's Reversal", Dawn, 15 March 2018.

48 2018 Opinion Piece English, Zahid Hussain, "Challenges to Democracy", Dawn, 23 May 2018.

rhetoric of change guised in democratic ideals can be classified as a 'post-truth' electoral discourse in Pakistan.

5 Conclusion

This chapter engages in a discussion on the operationalisation of the term 'post-truth' and using the case of Pakistani electoral discourse, specifically the discourse on democracy as an example of a 'post-truth' discourse. The shift in conceptualising post-truth in relation to discourse, and steering away from conflating post-truth with fake news broadens the scope of what constitutes as post-truth – it not misinformation, rather the elements of *lack*, *fantasy* and *emotionality* intertwine to depict a *phantasmatic logic* which renders the discourse 'post-truth'. The evaluation of a 'post-truth' discourse presented in this chapter highlights that for any political narratives to be constituted as 'post-truth', there is a need to reconceptualise the means in which 'post-truth' is identified in discourse.

Bibliography

Arendt, H., 1972. *Crisis Of The Republic*. New York: Harcourt.

Baggini, J., 2017. *A Short History Of Truth: Consolations For A Post-Truth World*. Quercus Publishing.

Barrera, O., Guriev, S., Henry, E. and Zhuravskaya, E., 2020. Facts, alternative facts, and fact checking in times of post-truth politics. *Journal of Public Economics*, 182.

Bilgin, P., 2017. Resisting Post-truth Politics, a Primer: Or, How Not to Think about Human Mobility and the Global Environment. *Global Policy*, 8.

Black, M., 1982. The Prevalence of Humbug. *Philosophic Exchange*, *13*(1).

d'Ancona, M., 2017. *Post-truth: The new war on truth and how to fight back*. Random House.

Davis, E., 2017. *Post-truth: Why we have reached peak bullshit and what we can do about it*. Little, Brown Book Group.

Davies, W., 2016. The age of post-truth politics. *The New York Times*, *24*.

Flynn, D.J., Nyhan, B. and Reifler, J., 2017. The nature and origins of misperceptions: Understanding false and unsupported beliefs about politics. *Political Psychology*, *38*, pp. 127–150.

Frankfurt, H.G., 2009. *On bullshit*. Princeton University Press.

Glynos, J. and Howarth, D., 2007. *Logics of critical explanation in social and political theory*. Routledge.

Glynos, J. and Stavrakakis, Y., 2008. Lacan and political subjectivity: Fantasy and enjoyment in psychoanalysis and political theory. *Subjectivity*, *24*(1).

Howarth, D., 2000. Discourse theory and political analysis. Manchester University Press.

Higgins, K., 2016. Post-truth: a guide for the perplexed. *Nature*, *540*(7631).

Hyvönen, A. E., 2018. Careless Speech: Conceptualizing Post-Truth Politics. *New Perspectives*, *26*(3), pp.31–55.

Keyes, R., 2004. *The post-truth era: Dishonesty and deception in contemporary life*. Macmillan.

Kuklinski, J.H., Quirk, P.J., Jerit, J., Schwieder, D. and Rich, R.F., 2000. Misinformation and the currency of democratic citizenship. *Journal of Politics*, *62*(3).

Laclau, E. and Mouffe, C., 2014. *Hegemony and socialist strategy: Towards a radical democratic politics*. Verso.

Lacan, J., 1977. Seminar XI: The four fundamental concepts of psychoanalysis. *Trans. Alan Sheridan. Ed. Jacques Alain-Miller*. New York: WW Norton and Company.

Lewandowsky, S., et al., 2012. Misinformation and its correction: Continued influence and successful debiasing. *Psychological science in the public interest*, *13*(3), pp.106–131.

McComiskey, B., 2017. *Post-truth rhetoric and composition*. University Press of Colorado.

McIntyre, L., 2018. *Post-truth*. MIT Press.

Nyhan, B. and Reifler, J., 2010. When corrections fail: The persistence of political misperceptions. *Political Behaviour*, *32*(2).

Nyhan, B. and Reifler, J., 2015. Displacing misinformation about events: An experimental test of causal corrections. *Journal of Experimental Political Science, 2*, pp. 81–93.

Orwell, G., 2017. *Orwell on Truth*. London: Harvill Secker.

Pepinsky, T.B., 2019. The return of the single-country study. *Annual Review of Political Science*, *22*, pp.187–203.

Seifert, C.M., 2017. The distributed influence of misinformation. *Journal of Applied Research in Memory and Cognition,* 6(4), pp. 397–400.

Shin, J., Jian, L., Driscoll, K. and Bar, F., 2018. The diffusion of misinformation on social media: Temporal pattern, message, and source. *Computers in Human Behaviour, 83*.

Žižek, S., 1989. *The sublime object of ideology*. Verso.

CHAPTER 9

A Challenge of Predictive Democracy: Rethinking the Value of Politics for the Era of Algorithmic Governance

Filip Biały

The advent of the current incarnation of artificial intelligence has raised a number of questions related to concepts and values such as freedom, legitimacy, agency or autonomy. While these issues are often presented as a challenge to human dignity or to the idea of humanity itself, they are not just of ethical or philosophical importance – they demand to be addressed as *political*. This is precisely the aim of this chapter: to develop a perspective that could help in thinking about the political implications of AI. The approach advocated here is based on the observation that political concepts form systems that structure political order and are integral parts of actual, real-world political thinking – of how people make sense of social reality in their everyday lives (Freeden, 2013). If meaning of the concepts changes, the coherence of the whole political order that relies on them might be questioned. For example, if human autonomy is reduced as a result of ceding decision-making powers to machines, we may start wondering about the legitimacy of democratic institutions. This is because thus far democratic legitimacy has been based on the assumption that the institutions are supported and governed by autonomous individuals. Such an idea is not only of theoretical importance. It is also a part of actual political thinking, so its deterioration may lead to real-world erosion of trust in political institutions, which, in turn, may affect the sense of political obligation and compliance. However, before we can make sense of these and similar dilemmas, we need a systematic procedure – a method for analysing conceptual change in the era of algorithmization of politics.

The analysis here will be limited to narrow AI, understood as a variety of forms of machine learning models and algorithms. While the prospects of possible artificial general intelligence or superintelligence have been widely discussed (Barrat, 2013; Bostrom, 2014; Tegmark, 2017) and criticised (Larson, 2021), the pervasiveness of narrow AI in politics suffices as a reason to conduct an investigation into its conceptual consequences. Accordingly, the focus of this chapter will be on using narrow AI within political decision-making that is an essential feature of the so-called algorithmic governance.

 | DOI:10.1163/9789004547261_011

What exactly happens to the conceptual system of democratic politics as a result of technological change? What is, politically speaking, at stake when we think about replacing human decision-making with the tools of algorithmic governance? How should we think about politics itself, when offered a promise of more efficient decisions? We will propose here that to position ourselves to respond to these and related questions, we could start with using a certain form of thought experiment: an intuition pump of predictive democracy – an idea of a system of ultimate, data-driven policymaking. In combination with teleological method of reasoning, it will allow for jump-starting much needed analysis of the conceptual shifts that results from the introduction of narrow AI into politics. Furthermore, although we will use an abstract thought experiment as a stepping stone, the overarching methodological aim is to elicit concrete, situated intuitions and convictions. As such, the method proposed here goes against the tendency to perceive datafication and algorithmization of contemporary societies as uniform processes, frequently described with implicit Western-centric bias. While technical solutions have some inherent properties, the results of an actual use of technology always depends on a particular socio-political context. Arguably, no meaningful conceptual analysis can be conducted without at least acknowledging the context sensitivity of political concepts.

Various forms of AI had been previously used in political decision-making, beginning with expert systems, introduced in the 1970s and developed in the 1980s (Schoech et al., 1985). A pivotal change occurred in the 2010s with the advent of deep learning (Mitchell, 2019: 63). The new technology was a result of a paradigm shift from symbolic AI, which was the basis of expert systems, towards connectionist approach, based on statistics, probability (Pearl, 1988) and artificial neural networks (Mead and Ismail, 1989). The availability of larger and more complex data (Big Data) as well as growing computing power led to not merely quantitative, but qualitative change in decision making (de Laat, 2018; Giest, 2017). While the rule-based experts systems of the 1980s were criticized as ineffective (e.g., Ascher, 1989; Weintraub, 1989), narrow AI may be deployed in a variety of contexts – frequently without sufficient scrutiny. Examples that have recently raised ethical questions include oft-discussed criminal sentencing (McKay, 2020), predictive policing (Shapiro, 2017), refugee resettlement (Bansak et al., 2018), education (Wang, 2021a, 2021b), healthcare (Reddy et al., 2020) and dealing with the COVID-19 pandemic (Vaishya et al., 2020). Because of its adaptability and flexibility, narrow AI might be, in principle, used to support, augment, or replace human decision-making in many other situations. In such a role, it may change the properties of political decision-making and of politics itself. That is why there is a vital need to investigate possible conceptual implications of algorithmic governance.

This introductory section is followed by four parts. It begins with a general overview of the recent literature on societal implications of AI with a conclusion that thus far the conceptual consequences of algorithmization of politics have not been comprehensively addressed. As a remedy, in the two following sections, the idea of predictive democracy will be introduced and then put to work to unearth and analyse intuitions about the concept of politics in the context of algorithmic governance. In the final section, this concept will be used to critically examine selected proposals of algorithmic governance.

1 Overlooked Challenges

Since its very beginnings the AI research has been accompanied by critical reflection (Borenstein et al., 2021), with interventions from some of the most important AI scholars (Weizenbaum, 1976; Wiener, 1950). Recently the introduction of deep learning and its public triumphs – such as the victory of AlphaGo over Lee Sedol in 2016 – witnessed a renewed interest in societal impacts of AI. Contemporary debates on the subject can be divided into three overlapping waves (Kind, 2020) or, more helpfully, into three connected subfields. The first one is the ethics of AI – a wide-ranging discourse on moral principles and consequences of AI in both academic literature as well as in numerous ethical codes and guidelines (Hagendorff, 2020; Powers and Ganascia, 2020). The second subfield is technical AI safety, dominated by computer scientists concerned about existential risks, resulting from the problem of control over both current and future AI (Juric et al., 2020; Russell, 2019). The third subfield, AI governance, focuses on the vast array of issues that require legislation (Wischmeyer and Rademacher, 2020), providing policy and research roadmaps (Dafoe, 2018) as well as regulatory proposals (European Commission, 2021).

What seems to be missing from these debates is a deeper political focus. Because of the pressure to quickly move from "mere" ethical considerations to regulation, political implications of AI have been discussed only marginally (cf. Hagendorff, 2020: 105; McQuillan, 2019: 163). While there is a growing number of empirical research related to the use and abuse of AI-driven tools in political communication (Woolley, 2016), campaigning and elections (Manheim and Kaplan, 2019), it has not yet amounted to any theoretical generalizations that would provide a reassessment of basic political concepts. Even in influential syntheses of datafication that offer interesting conceptual inventions – such as the idea of surveillance capitalism (Zuboff, 2019) – the analysis of existing concepts is lacking. This omission of the political dimension is also noticeable in the debate on the so-called value alignment problem (Christian, 2020). The problem, sometimes identified with the problem of control, is concerned with a question of how to make sure AI will comply with human values. Most of the attention is being given to the technical challenges: how to model and implement values into AI systems (Loreggia et al., 2020; Taylor et al., 2020)? The other, normative side of the problem is a question which (and whose) values should be implemented in a situation of ethical (and political) pluralism. We will readily observe that it is precisely the question political theorists have been asking with regard to political institutions in general: what kind of values should be selected as guiding principles by the state or other entities, and how the selection process should look like? Thus far there have been only few interventions that noticed this similarity (Gabriel, 2020).

The value alignment problem is an area that would benefit greatly from the input of political theory. However, because of its technical nature and an unhelpful tendency to pose it in the context of future forms of AI, it is only when we put the axiological questions in relation to actual decision-making, the most important conceptual implications come to the fore. That is why it seems that the most promising way to address these issues it to focus on the emerging research on the use of algorithms in politics. The research originated with concepts of algocracy and algocratic governance, first proposed in the context of studies on labour migration (Aneesh, 2006). Since then, other labels have been used to describe the phenomenon, such as algorithmic regulation (O'Reilly, 2013; Yeung and Lodge, 2019), government by algorithm (Engstrom et al., 2020) and algorithmic governance (Katzenbach and Ulbricht, 2019). It has been defined very broadly as a form of social ordering that results from using algorithms in public institutions (Yeung, 2018). In a more limited sense, it can be understood as the use of narrow AI algorithms to support, augment, or replace human decision-making. Such a definition of algorithmic governance helps in focusing on its perceived conceptual implications. Those will be made more visible as we turn to the idea of predictive democracy.

2 Intuition Pumps and AI Democracy

An analytic tool that we propose to use is intuition pump, which is a particular form of thought experiment (Dennett, 2014). The idea of intuition pumps was originally used in a pejorative sense to criticize John Searle's Chinese room thought experiment (Hofstadter and Dennett, 1988: 353–382). Since then, their role has been described more positively as "wonderful imagination grabbers, jungle gyms for the imagination" which "structure the way you think about a problem" (Dennett, 1996: 181–182). In fact, intuition pumps such as Thomas Hobbes's idea of the state of nature or John Rawls's veil of ignorance, have been for a long time used to advance philosophical arguments. They allow to focus on particular aspects of a problem by invoking intuitional responses that can be then analysed more systematically. For this second step we will use here a teleological method of reasoning that originated in Aristotelian ethics and has been used to resolve questions of justice (cf. Sandel, 2010: 188–190). Although teleology has been traditionally interested in discovering final moral ends, here it is understood simply as a way of asking and reasoning about a perceived purpose (or *telos*) ascribed to a concept within particular society.

An intuition pump offered here is the idea of predictive democracy. Let us imagine a political system in which all decision-making has been transferred into narrow AI-powered machines. The machines make decisions that are based on the preferences of the citizens. Those preferences are not collected through regular elections. They are instead inferred and predicted with the use of behavioural data on the individuals within the society. Since in predictive democracy politicians are no longer needed, the system does not know corruption that undermine stability of contemporary democracies. By getting rid of the representative layer of warring political parties, the decisions are made more swiftly, with direct relation to actual preferences of the people.

Although such a story might seem far-fetched, some versions of it have been invoked in recent debates. Jamie Susskind presents similar concepts of "data democracy" (Susskind, 2018: 246–250) that he ascribes to Hiroki Azuma (Azuma, 2014) and Yuval Noah Harari (Harari, 2017) as well the idea of "AI democracy" (Susskind, 2018: 250–254). César Hidalgo postulates an idea of "augmented democracy" in which personalized AI representatives (digital twins) would take part in making decisions (Hidalgo, 2019). Henrik Skaug Sætra, offering a "shallow defence of technocracy of artificial intelligence", claims that "it has the potential to revitalise democracy, increasing public participation in politics, and provide more efficient outcomes" (Sætra, 2020: 9). Peter Blook postulates abandoning anthropocentrism and embracing the possibility of trans-human democracy, based on "predictive politics" in which human and non-human beings would be entangled in a process of "mutual intelligent design" (Bloom, 2020: 173–210).

What is characteristic of those and similar proposals is that they treat political concepts like democracy and politics in an unspecific manner – or as doing the same conceptual work, regardless of context. That is why it is important to stress two further, methodological properties of the approach developed here. The first one is that the intuitions elicited by using intuition pumps, are bound to be biased or, rather, to be limited by context (cf. Brendel, 2004: 96). What some authors perceive as a weakness of intuition pumps or thought experiments in general (Brownlee and Stemplowska, 2017: 26), here is considered to be their strength. That is because one of the main criticisms that can be posed against technological solutionism – of which some ideas of algorithmic governance are instances – is its *caetextia* or inability to see the socio-political context. Any discussion on the implications of deploying algorithms into political decision-making should make clear that the outcomes will always be a resultant of a confrontation between a technological solution and pre-existing social, political and cultural circumstances. Similarly, the teleological part of the analysis should also be considered as context-sensitive, bearing different results for different societies. Teleology here is conceived not as a way to discover the universal nature or essence of a given concept, but to expose and unpack its cultural load. The proposed method should be thus considered as contextually-limited by design – or rather as open for an unlimited number of contexts, in which it may lead to differing conclusions.

The second property of our procedure is that intuition pumps are arbitrarily constructed mechanisms that always have several features that could be arranged differently, eliciting different intuitions. These features had been compared to "knobs" (Hofstadter and Dennett, 1988: 373–382). By turning the knobs it is possible to not only understand what an intuition pump and its parts are doing (Dennett, 2014: 79), but also to reconfigure the tool. The three main knobs in the idea of predictive democracy are the following: 1) a knob that regulates the range of decision-making: in predictive democracy it is assumed that *all* decision-making would be transferred into machines; 2) a knob that regulates the way people's preferences are collected: the assumption of predictive democracy is the preferences are inferred from behavioural data; 3) a knob that regulates the representative layer between the people and algorithmic decision-making: in predictive democracy there would not be human representatives. Each of these knobs can be set differently, resulting in different models of predictive democracy. Each of such models could bring out different kinds of intuitions about politics.

The use of intuition pump of predictive democracy, combined with teleological reasoning, offers a way to induce intuitions and initiate a context-sensitive analysis of political concepts. As such, the method is resistant to the notion of technological determinism. While it does not maintain that technology is a mere tool, it stresses that its effects should be considered against the background of existing social arrangements that influence actual technological uses. In that respect, the procedure is interested in making sense of how particular concepts and conceptual systems behave as a result of introducing specific technologies.

3 The Antinomies of Predictive Democracy

To properly understand the conceptual change that would result from establishing predictive democracy, we have to be reminded that interrelated political concepts structure both theoretical and real-world political thinking – and thus of political language of which they are building-blocks (Freeden, 2013: 31). Existing institutional settings may be considered an embodiment of a certain conceptual system. The idea of predictive democracy is, on the surface, such an imaginary institutional setting. But it is based on certain conceptual assumptions that we can now try to unearth.

Upon first hearing the story of predictive democracy, one may be struck by obvious differences between that idea and existing political arrangements. To a citizen of a Western democracy, with regularly held elections, parliamentary debates on legislative proposals between elected representatives, executive and judicial powers that implement and enforce enacted laws, the contrast with algorithmic decision-making is obvious. But what exactly is the conceptual difference here? Because political concepts are interconnected, we could start eliciting intuitions about any given concept and then proceed to those immediately adjacent or more remote ones. We could focus, for example, on the concept of representation and ask, what functions it performs that would be lacking in predictive democracy. Or we can try to understand, how important to democracy is a concept and a practice of elections. Better yet, we can ask about the concept of decision-making, trying to imagine the consequences of the difference between human and automated decision-makers. All these would be valid starting points of the analysis.

Here we will start with a concept that seems the most fundamental: the concept of politics. What understanding of politics is inscribed in the idea of predictive democracy? Because of the way the three knobs are set, the most immediate response is that politics seems to be reduced to decision-making. It is assumed that elections and political representation are just an outdated, inefficient way of collecting and transmitting preferences into decision-making process. By aligning decisions with preferences inferred from behavioural data, the algorithms would produce better results. An intuition here would be that politics in predictive democracy is about making decisions that would be based on the most accurate input, and thus produce the best possible output. How does it differ from the way we have understood politics thus far? We can now pose the same questions in a teleological manner: what do we consider to be the purpose of politics? We have been already given one idea: it is a good decision-making that itself may result in a good life, common good or happiness. But what are the other, competing or complementary aims of politics? We can come up with at least several of them, pointing out to different features of democratic political life. Politics, aside from decision-making, is about participation and deliberation, representation and belonging, agency and identity, as well as about compliance and contestation. While not all these elements might be considered *ultimate* goals of politics in themselves, their role is no less prominent than that of decision-making. Depending on a specific context, some of them may be more essential than others. Furthermore, because they are also political concepts that are mutually dependent, a redefinition or elimination of one or some of them would influence the whole system.

To get a better sense of how politics and other concepts would be redefined in predictive democracy, let us focus on the three knobs we have mentioned before. The first one is related to the range of decision-making power. The argument in favour of ceding such power to the machines is based on the assumption the algorithms are better than humans at making decisions. But what exactly do we mean by "better"? We may readily realize that most frequently the answer is better computational efficiency and optimization. That begs a question, whether all decisions can be made more efficient and optimized? If a decision is concerned with allocating resources, we may see some benefits of optimization. But when it comes to decisions on moral dilemmas – such as abortion, same-sex marriages or assisted suicide – it is difficult to apply the same principle. In these situations politics turns out to be about power and hegemony – about making decisions that cannot satisfy preferences of all sides. The aim of such decisions is rather to put a stop to political conflict or contest, without actually resolving the underlying tensions. These considerations mean that the proposed setting of decision-making knob would need important adjustments.

The second knob is responsible for the way in which people's political preferences are collected. The argument in favour of inferring preferences from behavioural data is that it would give a more accurate picture of the society and individuals. The problem with such a standpoint is that it seems to treat political preferences as ready to be predicted from the data. Intuitively we might feel that at least some of our political convictions do not pre-exist. They are rather a result of interactions with other citizens. In this respect, politics is about preference formation. Furthermore, the process of predicting political preferences indirectly, by correlating variables within the data on individuals, is diametrically different from the traditional way these preferences have been elicited, i.e. during elections or referendums. A closely connected question would be: what counts as a *political* preference? We can point out that many issues, related for example to sexual or gender, but also to racial identities, for a long time have not been considered political. It seems particularly important from the point of view of minorities whose needs and demands have been thus far ignored. For them, political participation is a way of making issues visible and of mobilizing support. On the one hand, it is probable that decision-making algorithms in predictive democracy would remain oblivious to such political dynamic and prove incapable of change. On the other hand, because of the limitation of the political sphere, predictive democracy could supress new demands and prevent new minority identities from being even constructed. Again, the settings of this knob require some serious rethinking as it is not entirely convincing that the preferences inferred from the data could be considered as the most accurate input. The notion of accuracy seems here rather misguided as predictive democracy would effectively change the way not only of how preferences, but also of how political subjectivities are formed.

Finally, we may attend to the third knob that regulates political representation which, in the case of predictive democracy, would be eliminated altogether. Why would we rely on inefficient representatives if algorithms would be better at making decisions? Do we need intermediaries when we have direct access to the data on citizens' political preferences? Aside from the responses given above, the answer could be that representation serves other functions or purposes. For example, being represented is connected with the sense of political belonging which, in turn, may affect the level of compliance. Furthermore, having human representatives that engage in political deliberation, gives the public a chance to listen to diverse opinions which is relevant to the preference-formation process discussed before. As with the previous two knobs, we need to tread carefully if we do not want to forfeit such features of politics.

This brief analysis is obviously preliminary and relies on arbitrary or contingent intuitions. However, even in this elementary form the reasoning shows something important not just about this version of algorithmic governance, but above all about certain elements of the concept of politics. The most general point is that while it is true that decision-making is an important aim of politics, there are other competing or complementing purposes ascribed to politics. This observation is followed by three more detailed points. First, even if we accept a decisionist conception of politics, political decision-making cannot be reduced to issues that aim at narrowly understood (or easily defined) efficiency. Second, politics has been a sphere of preference and identity formation and not just a tool to resolve tensions between already formed subjectivities. By changing the way preferences are collected, we would effectively change what we are collecting. Third, most of democratic politics today operates through representation which role is much more complex than just transmitting the preferences of the electorate into decision-making system. All this means that predictive democracy would pose profound challenges to the way politics and its various aspects has been conventionally understood in Western democracies. In the last section we will try to find out, how these challenges are handled by some of the proponents of algorithmic governance.

4 Is It All for the Better?

It may very well be argued that the idea of predictive democracy is extreme and unrealistic in its assumptions. Yet that is precisely the reason for constructing intuition pumps in the first place: they allow us to see more sharply how we understand our concepts. As a result, we are now better equipped to interrogate other proposals of algorithmic governance which can be considered variants of predictive democracy that have the above-mentioned knobs set a bit differently. By turning the knobs, one may come up with ideas of algorithmic governance better aligned with the requirements of the concept of politics that we have elicited in the previous section. This seems to be the case of the ideas of algorithmic governance mentioned before: data democracy, AI democracy, technocracy of AI and augmented democracy. Alternatively, one might try to offer a redefinition of politics that would accommodate technological change. In such case, the intuitions elicited by the idea of predictive democracy could serve as a reality check with regard to how far one can go (politically, if not theoretically) in subverting existing beliefs. In both cases, an important difference between the idea of predictive democracy and the proposals discussed below is that the former is a thought experiment, and the latter are offered as possible solutions to democracy's discontent.

On the most general level, although the proponents of the ideas for algorithmic governance may consider purposes of democratic politics other than decision-making, their discussion usually stops short of going into details of them. Susskind notes that citizens of data democracy would be deprived of conscious political participation that includes "noble actions" and allows people to flourish as humans. He admits that "[d]emocracy is about more than the competent administration of collective affairs" (Susskind, 2018: 249), but does not investigate this point further. While discussing the idea of AI democracy, he mentions that AI systems could remain subordinate to "traditional democratic processes like human deliberation and human votes" (Susskind, 2018: 253), again with no further explanation. Similarly, Sætra considers Aristotelian notion of humans as political animals as one of the objections to the technocracy of AI. In such a system, the objection goes, people cannot realise their political nature or their *telos*. Or, as the author phrases it, "[p]eople need full political participation to be satisfied" (Sætra, 2020: 5). His riposte is to claim that this is but one interpretation of Aristotle's ideas and that there are many other conceptions of the hierarchy of human activities. Another counterargument is that political participation today is already limited, and the notion of "political animal" is actually an argument against any form of non-direct democracy. Nevertheless, the author claims that technocracy of AI "may in fact lead to a revitalisation of popular participation and the accessibility of the political domain", through, among others, "algorithmic co-creation" that would give people "*more* political power" that they have today (Sætra, 2020: 5). On the same note, Hidalgo argues that introduction of digital twins in augmented democracy's would "empower people's ability to participate by technologically expanding their cognitive bandwidth" (Hidalgo, 2019). However, as with the previous proposals, this solution clearly perceives decision-making as central aim of politics.

In the case of the first of the three detailed points – related to the range of algorithmic decision-making – the quoted authors seems attracted to the promise of algorithmic efficiency. Characteristically, they invoke examples of AI superiority in making decisions in the sphere of economy. "If Deep Knowledge Ventures, a Hong-Kong based investor, can appoint an algorithm to its board of directors, is it so fanciful to consider that in the digital lifeworld we might appoint an AI system to the local water board or energy authority?", asks Susskind (Susskind, 2018: 251). Sætra bases his argument in favour of AI decision-making superiority on the idea of "AI economist" (Zheng et al., 2020), that operates on "reinforcement learning and multi-layered agent-based simulations could make better tax policies, improving *productivity* while simultaneously reducing income *inequality*" (Sætra, 2020: 1). How it would translate to decisions on moral issues? Susskind admits that data democracy could have a problem with resolving ethical issues as "[d]ata shows us what *is*, but it doesn't show what *ought to be*" (Susskind, 2018: 249). It means that prediction of algorithms might not be seen "as a substitute for moral reasoning", and such a system "would therefore need to be overlaid with some kind of overarching moral framework, perhaps itself the subject of democratic choice or deliberation", adding that "Data Democracy might be more useful at the level of policy rather than principle" (Susskind, 2018: 250). The author believes that in AI democracy the technological systems "could be made subject to the ethics of their human masters", so it "should not be necessary for citizens to surrender their moral judgement if they don't wish to" (Susskind, 2018: 253). We can see an important concession here: AI has its limitations in case of ethics or morality. That is not necessarily obvious to Sætra, who first phrases the moral objection in a rather peculiar way: "Computers should not make decisions that affect people's lives and wellbeing" (Sætra, 2020: 6). He then mentions several possible responses, from crowd-sourcing morality across the world and using it to train decision-making algorithms, to quantifying moral biases across cultures, claiming that it leads to machines that are becoming better at understanding human morality. The author does not address the problem of incommensurability of moral systems, nor does he explain, why globally crowd-sourced morality would be appropriate basis for resolving moral issues in local contexts. Furthermore, it is far from obvious that by being able to find some statistical patterns within the data on people's moral preferences, the machines gain any amount of understanding of morality. Admittedly, the author bases his position in favour of an AI technocracy on a premise that "[p]olicies should be evaluated on the basis of the fundamental moral values of the society in question, and ascertaining these values is the first purpose of politics" (Sætra, 2020: 7). Yet in the end he dismisses moral worries, claiming that not only AI, but "*all* technologies have moral and political *implication*" as they affect human wellbeing, and to get rid of them because of that would be absurd (Sætra, 2020: 8).

With regard to second point – politics as a sphere of preference formation – the proponents of algorithmic governance argue that it is possible to use technology to strengthen or revitalise the public sphere, allowing for more interaction between citizens. Nevertheless, they do not necessarily consider the role of political activity as a way of individual and collective preference formation. In fact, they seem to assume that people's preferences are ready to be extracted. Susskind claims that "one of the main purposes of democracy is to unleash the information and knowledge contained in people's minds and put it to political use", and subscribes to the idea that data – instead of, for example, explicitly stated political opinions – "reveals our actual lives and priorities" (Susskind, 2018: 246, 252). He goes on to argue that "more advanced model might involve the central government making inquiries of the population *thousands of times each day*, rather than once every few years – without having to disturb us at all" (Susskind, 2018: 252). Sætra discusses the importance of participation to democratic legitimacy and mentions the potential objection to the technocracy of AI that is "based on the idea that only democracy has the pedagogical effects ensuing from participation in politics" (Sætra, 2020: 6). He then points out to the already mentioned possibility of citizens' participation in creating algorithms, but does not consider what would be the implications of such a shift in terms of how the preferences are formed. In his more technically detailed account Hidalgo proposes that the information on people's preferences would be provided to citizens' digital twins in the form of both active and passive data. "Active data includes surveys, questionnaires, and, more importantly, the feedback that people would provide directly to its digital twin", while "[p]assive data would include data that users already generate for other purposes, such as their online reading habits (...), purchasing patterns, or social media behavior" (Hidalgo, 2019). The hope is that by allowing digital avatars supplied with such an information to vote on issues, people would be offered a better understanding of the questions and a way to conduct an informed debate as "[t]he idea of Augmented Democracy is never to offload the democratic process to a senate of digital twins but to empower people's ability to participate by technologically expanding their cognitive bandwidth" (Hidalgo, 2019). Nevertheless, such a solution would fundamentally change the way people interact with each other and how their preferences are formed.

Thirdly, we may interrogate the way algorithmic governance could resolve the problem of representation. Here again the most pronounced example is the idea of augmented democracy. While digital twins would not autonomously make decisions on people's behalf, they would act as a direct representation of each individual, replacing traditional representative mechanisms. That raises the question about the role of representation as a way of furthering issues related to collective, and especially minority identities. Such an individualistic bias could also be observed in Susskind proposals. His sense is that data-based AI democracy "would have a far greater claim to taking into account the interests of the population than the competitive elitist model with which we live today" (Susskind, 2018: 253). In his quest to remedy the shortcomings of "competitive elitist framework", the author does not consider the importance of representative model for non-individualistic political agency. Another argument for maintaining human representatives would be to exercise oversight over algorithmic decision-making systems, which only mentioned briefly by Sætra (2020: 5).

The discussion reveals several important blind spots in the ideas of algorithmic governance that have been trying to accommodate technological solutions to the requirements of contemporary politics. The alternative considered in recent literature is to abandon the traditional, anthropocentric understanding of politics and adopt the trans-human or post-human perspective (Bloom, 2020; Kalpokas, 2019) that perceives politics as a sphere in which humans and their artifacts are treated as an assemblage. Although we cannot go into details of these proposals here, it should be noted that such reformulations, as a way of theorising about new technological realities, are often persuasive. The most fundamental objection to them would concern their feasibility in making impact on the actual political thinking and, as the proponents of a more realistic or non-ideal theory would argue – in guiding political action (Valentini, 2012: 654). Another worry would be that an unreflective adoption of post-humanistic optics opens the door to human values being dominated by narrowly understood, technological principles of efficiency and optimization.

5 Conclusion

The challenge of predictive democracy has been offered here as a way to think about the political consequences of an already occurring technological shift. By eliciting contextually-sensitive intuitions and applying the teleological reasoning, it may be used to uncover our most immediate reactions to the prospects of algorithmic governance as well as to look critically at the proposals of rethinking politics for the era of artificial intelligence. In the end, we may find ourselves better prepared for the task of answering questions not only about our political concepts, but also about the value of politics itself. Although we have tried to present certain methodological procedure, this task – at least for the time being – cannot be easily algorithmized and demands human, and not artificial intelligence.

References

Aneesh, A. (2006). *Virtual Migration: The Programming of Globalization*. Durham University Press.

Ascher, W. (1989). Limits of "expert systems" for political-economic forecasting. *Technological Forecasting and Social Change*, *36*(1–2), 137–151. https://doi.org/10.1016/0040-1625(89)90019-X.

Azuma, H. (2014). *General Will 2.0: Rousseau, Freud, Google*. http://search.ebscohost.com/login.aspx?direct=true&scope=site&db=nlebk&db=nlabk&AN=852516.

Bansak, J. et al. (2018). *Improving refugee integration through data-driven algorithmic assignment. Science*, 359, 325–329. https://doi.org/10.1126/science.aao4408.

Barrat, J. (2013). *Our final invention: Artificial intelligence and the end of the human era*. Thomas Dunne Books.

Bloom, P. (2020). *Identity, Institutions and Governance in an AI World: Transhuman Relations*. Springer International Publishing. https://doi.org/10.1007/978-3-030-36181-5.

Borenstein, J. et al. (2021). AI Ethics: A Long History and a Recent Burst of Attention. *Computer*, *54*(1), 96–102. https://doi.org/10.1109/MC.2020.3034950.

Bostrom, N. (2014). *Superintelligence: Paths, dangers, strategies*. Oxford University Press.

Brendel, E. (2004). Intuition Pumps and the Proper Use of Thought Experiments. *Dialectica*, *58*(1), 89–108.

Brownlee, K., and Stemplowska, Z. (2017). Thought Experiments. In A. Blau (Ed.), *Methods in Analytical Political Theory* (1st ed.). Cambridge University Press. https://doi.org/10.1017/9781316162576.

Christian, B. (2020). *The Alignment Problem: Machine Learning and Human Values*. W.W. Norton & Company.

Dafoe, A. (2018). *AI Governance: A Research Agend*. Centre for the Governance of AI, Future of Humanity Institute, University of Oxford.

de Laat, P. B. (2018). Algorithmic Decision-Making Based on Machine Learning from Big Data: Can Transparency Restore Accountability? *Philosophy & Technology*, *31*(4), 525–541. https://doi.org/10.1007/s13347-017-0293-z.

Dennett, D. C. (1996). Intuition Pumps. In J. Brockman (Ed.), *The Third Culture: Beyond the Scientific Revolution*. Simon & Schuster.

Dennett, D. C. (2014). *Intuition Pumps and Other Tools for Thinking*.

Engstrom, D. F. et al. (2020). Government by Algorithm: Artificial Intelligence in Federal Administrative Agencies. *SSRN Electronic Journal*. https://doi.org/10.2139/ssrn.3551505.

European Commission. (2021). *Proposal for a regulation of the European Parliament and of the Council laying down harmonised rules on Artificial Intelligence (Artificial intelligence Act) and amending certain Union legislative acts*. European Commission.

Freeden, M. (2013). *The political theory of political thinking: The anatomy of a practice*. Oxford University Press.

Gabriel, I. (2020). Artificial Intelligence, Values and Alignment. *Minds and Machines*, *30*(3), 411–437. https://doi.org/10.1007/s11023-020-09539-2.

Giest, S. (2017). Big data for policymaking: Fad or fasttrack? *Policy Sciences*, *50*(3), 367–382. https://doi.org/10.1007/s11077-017-9293-1.

Hagendorff, T. (2020). The Ethics of AI Ethics: An Evaluation of Guidelines. *Minds and Machines*, *30*(1), 99–120. https://doi.org/10.1007/s11023-020-09517-8.

Harari, Y. N. (2017). *Homo Deus: A Brief History of Tomorrow*. Harper.

Hidalgo, C. (2019). *Augmented Democracy*. https://www.peopledemocracy.com/.

Hofstadter, D. R., and Dennett, D. C. (Eds.). (1988). *The mind's I: Fantasies and reflections on self and soul*. Bantam Books.

Juric, M., Sandic, A., and Brcic, M. (2020). AI safety: State of the field through quantitative lens. *2020 43rd International Convention on Information, Communication and Electronic Technology (MIPRO)*, 1254–1259. https://doi.org/10.23919/MIPRO48935.2020.9245153.

Kalpokas, I. (2019). *Algorithmic Governance: Politics and Law in the Post-Human Era*. Springer International Publishing. https://doi.org/10.1007/978-3-030-31922-9.

Katzenbach, C., and Ulbricht, L. (2019). Algorithmic governance. *Internet Policy Review, 8*(4). https://doi.org/10.14763/2019.4.1424.

Kind, C. (2020). The term 'ethical AI' is finally starting to mean something. *VentureBeat*. https://venturebeat.com/2020/08/23/the-term-ethical-ai-is-finally-starting-to-mean-something/.

Larson, E. J. (2021). *The myth of artificial intelligence: Why computers can't think the way we do*. The Belknap Press of Harvard University Press.

Loreggia, A. et al. (2020). Modeling and Reasoning with Preferences and Ethical Priorities in AI Systems. In S. M. Liao (Ed.), *Ethics of Artificial Intelligence*. Oxford University Publication.

Manheim, K. and Kaplan, L. (2019). Artificial Intelligence: Risks to Privacy and Democracy. *Yale Journal of Law & Technology, 21*(1), 106+. Gale Academic OneFile.

McKay, C. (2020). Predicting risk in criminal procedure: Actuarial tools, algorithms, AI and judicial decision-making. *Current Issues in Criminal Justice, 32*(1), 22–39. https://doi.org/10.1080/10345329.2019.1658694.

McQuillan, D. (2019). The Political Affinities of AI. In A. Sudmann (Ed.), *The Democratization of Artificial Intelligence: Net Politics in the Era of Learning Algorithms*.

Mead, C. and Ismail, M. (Eds.). (1989). *Analog VLSI Implementation of Neural Systems*.

Mitchell, M. (2019). *Artificial Intelligence: A Guide for Thinking Humans*.

O'Reilly, T. (2013). Open Data and Algorithmic Regulation. In B. Goldstein and L. Dyson (Eds.), *Beyond transparency: Open data and the future of civic innovation*. Code for America Press.

Pearl, J. (1988). *Probabilistic reasoning in intelligent systems: Networks of plausible inference*. Kaufmann.

Powers, T. M., and Ganascia, J.-G. (2020). The Ethics of the Ethics of AI. In M. D. Dubber, F. Pasquale, and S. Das (Eds.), *The Oxford Handbook of Ethics of AI* (pp. 25–51). Oxford University Press. https://doi.org/10.1093/oxfordhb/9780190067397.013.2.

Reddy, S. et al. (2020). A governance model for the application of AI in health care. *Journal of the American Medical Informatics Association, 27*(3), 491–497. https://doi.org/10.1093/jamia/ocz192.

Russell, S. J. (2019). *Human compatible: Artificial intelligence and the problem of control*. Viking.

Sætra, H. S. (2020). A shallow defence of a technocracy of artificial intelligence: Examining the political harms of algorithmic governance in the domain of government. *Technology in Society, 62*, 101283. https://doi.org/10.1016/j.techsoc.2020.101283.

Sandel, M. J. (2010). *Justice: What's the right thing to do?* (1st pbk. ed). Farrar, Straus and Giroux.

Schoech, D. et al. (1985). Expert Systems: Artificial Intelligence for Professional Decisions. *Computers in Human Services, 1*(1), 81–115. https://doi.org/10.1300/J407v01n01_06.

Shapiro, A. (2017). Reform predictive policing. *Nature, 541*(7638), 458–460. https://doi.org/10.1038/541458a.

Susskind, J. (2018). *Future Politics: Living Together in a World Transformed by Tech.* Oxford University Press.

Taylor, J. et al. (2020). Alignment for Advanced Machine Learning Systems. In S. M. Liao (Ed.), *Ethics of Artificial Intelligence.* Oxford University Publication.

Tegmark, M. (2017). *Life 3.0: Being human in the age of artificial intelligence.* Alfred A. Knopf.

Vaishya, R. et al. (2020). Artificial Intelligence (AI) applications for COVID-19 pandemic. *Diabetes & Metabolic Syndrome: Clinical Research & Reviews, 14*(4), 337–339. https://doi.org/10.1016/j.dsx.2020.04.012.

Valentini, L. (2012). Ideal vs. Non-ideal Theory: A Conceptual Map: Ideal vs Non-ideal Theory. *Philosophy Compass, 7*(9), 654–664. https://doi.org/10.1111/j.1747-9991.2012.00500.x.

Wang, Y. (2021a). When artificial intelligence meets educational leaders' data-informed decision-making: A cautionary tale. *Studies in Educational Evaluation, 69*, 100872. https://doi.org/10.1016/j.stueduc.2020.100872.

Wang, Y. (2021b). Artificial intelligence in educational leadership: A symbiotic role of human-artificial intelligence decision-making. *Journal of Educational Administration, 59*(3), 256–270. https://doi.org/10.1108/JEA-10-2020-0216.

Weintraub, J. (1989). Expert Systems in Government Administration. *AI Magazine, 10*(1), 3.

Weizenbaum, J. (1976). *Computer power and human reason: From judgment to calculation.* W. H. Freeman.

Wiener, N. (1950). *The Human Use of Human Beings; Cybernetics and Society.*

Wischmeyer, T. and Rademacher, T. (Eds.). (2020). *Regulating Artificial Intelligence.* Springer.

Woolley, S. C. (2016). Political Communication, Computational Propaganda, and Autonomous Agents – Introduction. *International Journal of Communication*, 9.

Yeung, K. (2018). Algorithmic regulation: A critical interrogation: Algorithmic Regulation. *Regulation & Governance, 12*(4), 505–523. https://doi.org/10.1111/rego.12158.

Yeung, K., and Lodge, M. (Eds.). (2019). *Algorithmic regulation.* Oxford University Press.

Zheng, S. et al. (2020). The AI Economist: Improving Equality and Productivity with AI-Driven Tax Policies. *ArXiv:2004.13332 [Cs, Econ, q-Fin, Stat]*. http://arxiv.org/abs/2004.13332.

Zuboff, S. (2019). *The age of surveillance capitalism: The fight for a human future at the new frontier of power.* PublicAffairs.

CHAPTER 10

Value Problems and Practical Challenges of Contemporary Digital Technologies

Julija Kalpokienė

It is beyond questioning that today's digital technologies, such as AI, bring fundamental changes and challenges to everyday life. Whether it is a matter of employment, communication, politics, personal relationships, or any other matter, there is usually some structuring and optimising digital process in place. Therefore, it is by no means surprising that there are growing calls for regulation of AI and similar technologies.

Nevertheless, such calls do not translate into an agreement over the substance and even extent of such regulation. In fact, positions on this matter are almost as numerous as the calls themselves. Therefore, this chapter aims to bring some order and coherence to the debate by way of reviewing some of the most characteristic propositions for how (not) to regulate AI while also proposing new ways of reframing the debate in light of the contributions to this book.

The overall picture that emerges is one of divergent paths whereby consensus over regulation is hard to come by while its extent and substance (if regulation is embraced at all) are not clear either. It is, therefore, one of the important contributions of this volume that without striving for any explicit synthesis (and without necessitating one as a structural precondition), the authors nevertheless point towards potential instances of dialogue and interconnection even among diverse concerns and points of view.

1 Approaching Regulation: Ethical and Technological Debates

Recent years have been marked by not just a growing debate around AI regulation but also by national and regional efforts to put such regulation into practice. Within the civil society and in some policy quarters there seems to be a significant drive towards self-regulation and making sure that AI is both responsible and trustworthy (de Laat 2021). An increasing number of proposals for national or regional AI supervision and certification agencies are put forward as well (Pomares and Abdala 2020). Similarly, Candelon et al. (2021) note

 | DOI:10.1163/9789004547261_012

a shift of regulatory focus in line with changing concerns of both citizens and politicians: from data protection and regulation to AI regulation (Candelon et al., 2021). On the other hand, it might be somewhat ambitious to claim, as e.g. Smuha (2021) does, that while states have, for quite a while already, been rushing to adopt AI technologies as soon and as extensively as possible in hope of obtaining technological and economic advantage, more recently one could witness a shift from 'race to AI' to 'race to AI regulation' in order to bring about greater trust and legal certainty. After all, existing efforts are too few and far between to merit claims of representing a 'race' to regulation. Nevertheless, it must be admitted that even if such a race would pick up, the interdependence of the global economy and the nature of the internet means that it could merely become a 'race to the bottom' as states and regional bodies compete for innovation and investment, leaving populations inadequately protected against potential risks of AI (Smuha 2021). Hence, any progress on the matter is far from guaranteed.

Technological change and digital innovation are sometimes seen as an 'earthquake' or 'giant bulldozer' that governments have only haphazardly tried to stay in control of (Wadhwa and Wadhwa 2021). Similarly, for Dempsey (2021), regulation is falling significantly behind innovation. Hence, it comes as no surprise that despite a growing attention to ethics and regulation, the trend is still of AI's implementation in potentially sensitive areas – from education to healthcare to public administration – without there being adequate safeguards in place (Pomares and Abdala 2020: 91). Nevertheless, regulation is necessary due to the impact that AI has on everyday lives of individuals in private, public, and business settings while being (on its own or in terms of the people behind it) unaccountable (Galaski 2021). Indeed, AI's impact is felt across wide-ranging manifestations from helping make significant policy decisions to disrupting the labour market (Ferretti 2021). Unsurprisingly, regulation advocates are calling for human rights to be kept at the heart of the regulation process, potentially including a ban on technologies that violate fundamental rights through their operation (Galaski 2021). Indeed, there is a paradox in terms of the ubiquity and importance of use on the one hand and a lack of transparency and accountability, particularly when bias occurs, on the other (Pomares and Abdala 2020). For this reason, despite there having been an overall shift over the past decades towards shrinking public regulation and increasing private (self-)governance, states are still obliged to act on AI regulation to ensure the fundamental rights of their citizens (Hoffmann-Riem 2020).

Focus on AI ethics is often encouraged as the missing link that mediates between centralised government regulation and industry self-regulation or corporate standard-setting to ensure that AI is a force for good (see e.g. Taddeo and Floridi 2018). Nevertheless, there simultaneously are fears that such ethics-first (or perhaps even ethics-only) approaches could lead to cherry-picking among norms and ethical frameworks in order to continue with business as usual inasmuch as possible – in short, these are issues of 'ethics washing' and trust in the AI industry (de Laat 2021). Moreover, there is also a danger of instrumentalising ethics by turning ethical frameworks into mere industry-serving tools to increase public acceptance and diminish the pressure for government regulation; a characteristic example here would be Taddeo and Floridi's (2018) argument that the primary need for AI ethics is simply to avoid societal rejection of AI and to prevent the loss of innovation potential that would allegedly come with such rejection. In this line of argument, AI is seen as inherently beneficial and, therefore, better left undisturbed with ethics paradoxically acting as a shield to defend AI from the easily excitable masses. It is thus unsurprising that, for others, talk of ethics merely represents an 'easy' or 'soft' solution when proper legal regulation is difficult or impossible to agree upon or even as a way to avoid regulation (see e.g. Wagner 2018). In a similar manner, for Hoffmann-Riem (2020), focus on AI ethics can even become potentially dangerous due to the risk of ending up with vague and non-binding principles without any real effect.

Simultaneously, one can observe a rise in the number of public-private initiatives concerned with AI ethics, pointing towards the debate becoming a fashionable one in governance, business, and academic circles (Mittelstadt 2019). Certainly, despite the doubts raised above, it would be futile to ignore the importance of governance based on ethics and other normative principles as such, particularly as privacy, transparency, fairness, interpretability and explainability, and allowing for ethics auditing are all features of crucial importance (Cath 2018: 2; see also Hoffmann-Riem 2020; Wulf and Seizov 2020). Likewise, the OECD (2019) stresses, in addition to the preceding, values such as sustainability, rule of law, security, and accountability. No less important are preservation of constitutional values and human rights, prevention of manipulation, ensuring fair economic development, etc. (Hoffmann-Riem 2020). But for such benefits to materialise, ethics *and* law should positively reinforce each other instead of ethics being left on its own (Wagner 2018). Notably, some would also argue that the inclusion of state law as a means to enforce ethics has normative significance: according to Ferretti (2021), regulation by governments is preferable because it allows for democratic vetting of norms with the onus being on technology companies to cooperate with governments in helping to establish adequate regulation.

Moreover, a realistic ethical framework for regulation must adopt an embedded approach to the extent that AI's ethical challenges are not mere design flaws but, instead, a reflection of the broader ethical, cultural, social, and political challenges inherent to contemporary societies; for this reason, AI ethics, as a regulatory endeavour, should be seen as a matter of ever-ongoing process and not as a matter of a few quick technological fixes (Mittelstadt 2019). But even then, governance by way of stipulating values may not bring the desired results. For example, transparency and explainability are limited in their effectiveness because AI tools are capable of learning and evolving on their own (Chesterman 2021). For this reason, more complex solutions may be sought.

There are also other kinds of calls for soft versions of AI governance – for example, Gasser and Almeida (2017) put forward an argument for a multi-layered and distributed system of AI governance, reminiscent of the current governance of the Internet. Others, meanwhile, would suggest embracing technical solutions for fixing what are presented as mere technical problems, such as creating 'fair' AI capable of evaluating its own biases (Feuerriegel, Dolata and Schwabe 2020). Potentially, one could also envisage as a solution AI learning to regulate *itself* by analysing the applicable laws (Chesterman 2021). This would, of course, give a completely new meaning to the idea of self-regulation. In a somewhat similar manner, Wiggers (2021) describes an already growing marketplace for outsourcing AI governance to purpose-made software (Wiggers 2021). Partly related to such solutions of giving autonomy to technological artefacts are suggestions of conferring legal personality to algorithms that are capable of developing autonomously beyond their creators' intentions, including the possibility of courts mandating the execution of a 'kill switch' (Treleaven, Barnett and Koshiyama 2019; similarly, see also Schirmer 2020).

2 Debating Regulation: the Feasibility and Desirability of New Laws

Across the present debates, it is not uncommon for regulation to be opposed in principle, claiming that there is no need for a general AI regulatory system – simply applying existing laws to AI would do (Reed 2018). In the same vein, Edelman (2020) warns against regulating AI as a standalone technology, instead proposing to treat it as merely a multifaceted tool. Seen in this light, focus should be on vetting procedures instead of lofty ethical and regulatory frameworks; should this kind of framework be adopted, omnibus legislation would be simply distractive – instead, it is claimed, it would be better to focus on narrow problem-and sector-specific issues without ascribing too much novelty and independence to AI (Edelman 2020). Elsewhere, it is suggested that application of a single branch of law might suffice to reduce many of the truly impactful challenges raised by AI – for example, Scheurer (2021) puts forward competition law as a candidate.

Indeed, lawmakers are forced to navigate the dire straits between the Scilla and Charybdis of AI regulation: precaution and innovation (Chesterman 2021). As digital innovations are highly mobile, today's AI landscape is characterised by global competition and global movement of technology; therefore, it is claimed, just as investors choose more favourable tax regimes over less favourable ones, innovators choose environments that are less burdensome in terms of regulation (Eggers, Turley and Kishnani 2018). Hence, it is not uncommon to encounter warnings of high cost and loss of innovation if regulation is embraced (see e.g. Bertuzzi 2021). In this case, regulation is seen to act either as a catalyst or as a hindrance – or, to be more precise, to act as a hindrance when present and as a catalyst when it is absent (Eggers, Turley and Kishnani 2018). On the other hand, too much caution is seen as dangerous because doing nothing or too little might fail to clear path for emerging disruptive technologies – although here again caution is typically seen as removing existing limitations (Fenwick, Kaal and Vermeulen 2017) and not as failure to create new rules for enhancing protection of citizens. Similarly, for Falzone (2012: 107), '[g]overnment is often at its best when it cultivates new technologies and then gets out of the way'. Nevertheless, there is also another facet of regulation for the benefit of innovation because without regulation the existing big players are able to stifle competition by setting the standards, acting as gatekeepers, and possessing the ability to outspend others on research and development (Dempsey 2021). Moreover, the presence of regulation can also bring clarity to the business environment, and, as a result, regulation and innovation are not mutually exclusive (Uppington 2021).

One more matter that is rather intensely debated is the extent to which technology companies themselves should be involved in AI regulation. For some, it is a matter of government independence in standard setting, which is seen as being jeopardised by technology companies often acting as semi-co-regulators, often themselves involved in setting or proposing regulatory standards (Cath 2018: 4). However, 'traditional' regulation by politicians can be seen as too slow and cumbersome, necessitating new modes of regulation involving industry representatives and big data – generated, of course, by the industry itself (Fenwick, Kaal and Vermeulen 2017). In other cases, meanwhile, cooperation (even if sometimes on a stick-and-carrot basis) is seen as crucial to effectiveness – in this case, it is claimed, a combination of self-regulation and credible government threats of regulation might work in terms of setting and upholding standards for the technology industry (Cusumano, Gower and Yoffie 2021).

The EU currently stands at the forefront of AI regulation with its efforts to establish a risk-based model of regulating for particular applications of AI (MacCarthy and Propp 2021). Due to its role as a trailblazer, the EU is likely to set the standard for AI regulation in the democratic world (Galaski 2021;

see also Knight 2021). Moreover, given the prominence of the EU market, its regulatory attempts will likely have global ramifications (Kahn 2021). However, the EU is not alone in this domain, with organisations such as OECD and the World Economic Forum having issued guidelines on AI, albeit non-binding ones (Galaski 2021). Nevertheless, The United Nations remains a glaring omission in the AI governance landscape, not just in terms of failing to act as a norms' entrepreneur but also in terms of not taking stock over its own AI use (Fournier-Tombs 2021). Meanwhile, among the first national level AI policies, one can witness prioritisation of ethics and 'human-centricity' in the UK and Singapore whereas the Chinese approach transpires to favour innovation and development while paying less attention to safeguarding humans (Murgia and Shrikanth 2019).

Still, regardless of what kind of regulation (if any) is embraced, harmonisation is key, particularly due to the challenges posed by differences in regional frameworks and divergent thresholds for meeting key requirements, such as explainability (Candelon et al., 2021). Indeed, there is a clear mismatch between the borderless nature of digital technologies and their fragmented regulation (Pomares and Abdala 2020; see also Hoffmann-Riem 2020). Similarly, for Chesterman (2021), at least as long as regulation, in its conventional form, still involves states, AI is difficult for individual states to regulate on their own, implying the necessity of international cooperation. Indeed, an argument could be made that the current patchwork of existing laws and industry self-regulation is inadequate and there is a need to build innovation-friendly overarching regulatory structures (Quest and Charrie 2019); however, such openness to innovation cannot come at the cost of reduced protection of citizens. Crucially, though, there is also a tension between national and supranational levels: regional or even global regulation might overlook national circumstances and specificities while a regulatory regime that is overly fractionalised would only stifle innovation, create tensions, and fail to ensure equal respect for human rights (Pomares and Abdala 2020: 87). Here again, it transpires, straightforward answers are elusive.

3 What Have We Learned So Far?

In this book's opening chapter, Kalpokas stresses the unavoidable mutual embeddedness and inseparability of the human and the technological, leading to the necessity to discard the anthropocentric illusion of human exceptionality and dominance. This observation is crucial to the effort of understanding the direction that the governance and regulation of technology should take.

After all, the rejection of anthropocentrism does not mean that regulation is not necessary – on the contrary, it only exemplifies the centrality of the ethical debate in order to find a mutually equitable framework for the interaction between humans and their (technological) environment. Meanwhile, Marsili reminds that human-AI interaction can easily become a matter of life and death. As military applications of AI can, due to their strategic and national security implications, easily become a race to the lowest common denominator, exploring ways to apply existing norms, such as those pertaining to International Humanitarian Law, is crucial. In this context, it is important to call, according to Marsili, for an interdisciplinary framework that cuts deep into the complexities of military AI. Nevertheless, Poseliuzhna's chapter provides some necessary counterweight: while the focus is typically on technologies, with sophisticated tools and innovation taken as a given, it is often forgotten that humans usually still perform a key and irreplaceable role in the development of new technologies. For this reason, competition over top talent is a vital part of the global AI competition, putting a dent into the argument that the technology world is highly mobile and reactive to regulatory changes; uneven distribution of talents means that some places will remain privileged at the expense of others regardless of other factors, such as differences in regulatory regimes.

Meanwhile, with growing applications of AI, questions of responsibility and liability for the damage caused by AI are going to become particularly pressing. In this case, de Bruyne's proposal for third-party certification points towards one of the possible ways to narrow the gap between regulation sceptics and enthusiasts by helping ensure the safety of citizens but simultaneously avoiding excessive and cumbersome legislation. In a similar manner, Petyk shows how the increase of communication and data transmission speeds, coupled with AI, acts to endanger certain fundamental rights, such as privacy. For this reason, it is important to highlight, as Petyk does, the importance of adherence to the highest security standards and protocols as well as to existing legislation and the responsibilities held by network providers. Simultaneously, the vulnerability of users is laid bare, strengthening the case for legislative intervention.

While others adopt a more speculative approach, Couzigou analyses the application and effectiveness of a real-life legislative intervention – the French anti-disinformation legislation. While the aim of the legislation is noble, its practical application transpires not to be without challenges. Ultimately, it is transparency and information literacy that are seen as key to fighting disinformation. This only comes to show that regulation on its own is not going to become the ultimate solution – instead, it needs broader social and institutional support. In a rather similar fashion, Barker discusses self-regulation as a deceptively simple solution: while it is easy to pass the buck to the industry itself (and it might also seem to be a less intrusive option), it all boils down to the extent to which the public can trust the process (and the industry itself). Hence, it is commendable that Barker stresses the importance of building trust (and the role of regulators in creating the conditions that facilitate this process) so that the public can feel secure. This only further underscores the necessity of taking the broader context into account when considering any regulatory options.

Even more explicitly than in the previous chapters, Kuraishi's starting point is not technology but the human as she ends up replacing the technology-centric account of post-truth with one deeply rooted in human characteristics. This only comes to show that for any regulatory endeavour to be successful, technology cannot be taken in isolation. Instead, the way in which humans make sense of the world and carve their space within it should be given attention. Likewise, Biały underscores a very similar point, albeit in a different way. His focus is on a critique of the application of optimisation strategies characteristic to machine learning and artificial intelligence in processes of democratic governance. In this way, technological governance of humans is shown to be a futile and self-defeating strategy that does not reflect the way in which human life is truly lived. Here one can close the circle by bringing Kalpokas' chapter back into consideration. Seen from this perspective, governance and regulation should then take place at the intersection of the human and the technological as opposed to prioritising either.

4 Conclusion

The existent and emerging digital autonomous technologies raise numerous questions as to their impact on humans in terms of their rights and freedoms, economic and political participation, and the very idea of what it means to be human. Unsurprisingly, the regulation of technologies, such as AI, is fiercely debated. The balance between ethics-and self-regulation-based approaches on the one hand and more stringent regulation through law on the other as well as the debate between regulation sceptics and optimists remain far from being solved, meaning that clarity and stability of the applicable regimes is hard to foresee. Still, the contributions to this book aim to cut across this polarised landscape by showing the necessity of more complex and nuanced solutions to the regulatory antinomies of today.

Bibliography

Bertuzzi, L. (2021, July 26). Study Warns of Compliance Costs for Regulating Artificial Intelligence. *Euractiv*, https://www.euractiv.com/section/digital/news/study-warns-of-compliance-costs-for-regulating-artificial-intelligence/.

Candelon, F. et al. (2021). AI Regulation is Coming: How to Prepare for the Inevitable. *Harvard Business Review*, https://hbr.org/2021/09/ai-regulation-is-coming.

Cath C. (2018). Governing Artificial Intelligence: Ethical, Legal and Technical Opportunities and Challenges. *Philosophical Transactions of the Royal Society A*, 376, 1–8.

Chesterman, S. (2021). *We, the Robots? Regulating Artificial Intelligence and the Limits of the Law*. Cambridge and New York: Cambridge University Press.

Cusumano, M. A., Gawer, A. and Yoffie, D. B. (2021). Can Self-Regulation Save Digital Platforms? *Industrial and Corporate Change*, doi: 10.1093/icc/dtab052.

de Laat, P. B. (2021). Companies Committed to Responsible AI: From Princiles Towards Implementation and Regulation? *Philosophy and Technology*, doi: 10.1007/s13347-021-00474-3.

Dempsey, P. (2021, August 16). Regulation for the AI-Driven New Age. *Engineering and Technology*, https://eandt.theiet.org/content/articles/2021/08/regulation-for-the-ai-driven-new-age/.

Edelman, R. D. (2020, January 13). Here's How to Regulate Artificial Intelligence Properly. *The Washington Post*, https://www.washingtonpost.com/outlook/2020/01/13/heres-how-regulate-artificial-intelligence-properly/.

Eggers, W. D., Turley, M. and Kishnani, P. K. (2018, June 19). The Future of Regulation: Principles for Regulating Emerging Technologies. *Deloitte*, https://www2.deloitte.com/us/en/insights/industry/public-sector/future-of-regulation/regulating-emerging-technology.html.

Falzone, A. (2012). Regulation and Technology. *Harvard Journal of Law and Public Policy*, 36(1), 105–107.

Fenwick, M. D., Kaal, W. A. and Vermeulen, E. P. M. (2017). Regulation Tomorrow: What Happens When Technology Is Faster than the Law? *American University Business Law Review*, 6(3), 561–594.

Ferretti, T. (2021). An Institutionalist Approach to AI Ethics: Justifying the Priority of Government Regulation over Self-Regulation. *Moral Philosophy and Politics*, doi: 10.1515/mopp-2020-0056.

Feuerriegel, S., Dolata, M. and Schwabe, G. (2020). Fair AI: Challenges and Opportunities. *Business & Information Systems Engineering*, 62, 379–384.

Fournier-Tombs, E. (2021, May 31). The United Nations Needs to Start Regulating the 'Wild West' of Artificial Intelligence. *The Conversation*, https://theconversation.com/the-united-nations-needs-to-start-regulating-the-wild-west-of-artificial-intelligence-161257.

Galaski, J. (2021, September 8). AI Regulation: Present Situation and Future Possibilities. *Liberties*, https://www.liberties.eu/en/stories/ai-regulation/43740.

Gasser, U. and Almeida, V. A. F. (2017). A Layered Model for AI Governance. *IEEE Internet Computing*, 21(6), 58–62.

Hoffmann-Riem, W. (2020). Artificial Intelligence as a Challenge for Law and Regulation. In T. Wischmeyer and T. Rademacher (eds.), *Regulating Artificial Intelligence*. Cham: Springer, pp. 1–25.

Kahn, J. (2021, April 21). Europe Proposes Strict AI regulation Likely to Have an IMpacr around the World. *Fortune*, https://fortune.com/2021/04/21/europe-artificial-intelligence-regulation-global-impact-google-facebook-ibm/.

Knight, W. (2021, April 21). Europe's Proposed Limits on AI Would Have Global Consequences. *Wired*, https://www.wired.com/story/europes-proposed-limits-ai-global-consequences/.

MacCarthy, M. and Propp, K. (2021, May 4). Machines Learn that Brussels Writes the Rules: The EU's New AI Regulation. *The Brookings Institution*, https://www.brookings.edu/blog/techtank/2021/05/04/machines-learn-that-brussels-writes-the-rules-the-eus-new-ai-regulation/.

Mittelstadt, B. (2019). Principles Alone Cannot Guarantee Ethical AI. *Nature Machine Intelligence*, 1, 501–507.

Murgia, M. and Shrikanth, S. (2019, May 30). How Governments Are Beginning to Regulate AI. *Financial Times*, https://www.ft.com/content/025315e8-7e4d-11e9-81d2-f785092ab560.

OECD (2019, May 22). Forty-two Countries Adopt New OECD Principles on Artificial Intelligence, https://www.oecd.org/going-digital/forty-two-countries-adopt-new-oecd-principles-on-artificial-intelligence.htm.

Pomares, J. and Abdala, M. B. (2020). The Future of AI Governance: The G20's Role and the Challenge of Moving beyond Principles. *Global Solutions Journal*, 5, 84–94.

Quest, L. and Charrie, A. (2019, September 19). The Right Way to Regulate the Tech Industry. *MIT Sloan Management Review*, https://sloanreview.mit.edu/article/the-right-way-to-regulate-the-tech-industry/.

Reed, C. (2018). How Should We Regulate Artificial Intelligence? *Philosophical Proceedings of the Royal Society A*, 376, 1–12.

Scheurer, S. (2021). Artificial Intelligence and Unfair Competition – Unveiling an Underestimated Building Block of the AI Regulation Landscape. *GRUR International*, 70(9), 834–845.

Schirmer, J. E. (2020). Artificial Intelligence and Legal Personality: Introducing "Teilrechtsfähigkeit": A Partial Legal Status Made in Germany. In T. Wischmeyer and T. Rademacher (eds.), *Regulating Artificial Intelligence*. Cham: Springer, pp. 123–142.

Smuha, N. A. (2021). From a 'Race to AI' to a 'Race to AI Regulation': Regulatory Competition for Artificial Intelligence. *Law, Innovation and Technology*, 13(1), 51–84.

Taddeo, M. and Floridi, L. (2018). How AI Can Be a Force for Good. *Science*, 361, 751–752.

Treleaven, P., Barnett, J. and Koshiyama, A. (2019). Algorithms: Law and Regulations. *IEEE Computer*, 52(2), 32–40.

Uppington, W. (2021, October 6). Driving AI Innovation in Tandem with Innovation. *TechCrunch*, https://techcrunch.com/2021/10/06/driving-ai-innovation-in-tandem-with-regulation/.

Wadhwa, V. and Wadhwa, T. (2021, March 1). The World Must Regulate Tech Before It's Too Late. *Foreign Policy*, https://foreignpolicy.com/2021/03/01/technology-ethics-regulation-dangers-global-consensus-cooperation/.

Wagner, B. (2018). Ethics as an Escape from Regulation. From "Ethics-Washing" to Ethics-Shopping? In E. Bayamlioglu et al. (eds.), *Being Profiled: Cogitas Ergo Sum*. Amsterdam: Amsterdam University Press, pp. 84–89.

Wiggers, K. (2021, August 25). How New Regulation is Driving the AI Governance Market. *VentureBeat*, https://venturebeat.com/2021/08/25/how-new-regulation-is-driving-the-ai-governance-market/.

Wulf, A. J. and Seizov, O. (2020). Artificial Intelligence and Transparency: A Blueprint for Improving the Regulation of AI Applications in the EU. *European Business Law Review*, 31(4), 611–640.

Index

Printed in the United States
by Baker & Taylor Publisher Services